An Urban Approach to Climate-Sensitive Design

Rapid urbanization in the tropics has brought in its wake many hitherto unknown changes to humans, animals and plants and the physical environment. Many of these changes are well studied by researchers in diverse fields such as medicine, agriculture and engineering. However, the climatic effect of urbanization has not been understood nor the knowledge base applied by urban designers, planners, architects and engineers. In particular, the energy and bio-climatic implications of changes induced by urbanization have received very little attention in urban design and planning. This book lays out the problem of tropical urban climate anomaly and points to possiblities of mitigating these changes through design and planning options.

An Urban Approach to Climate-Sensitive Design brings together the emerging literature on climate-sensitive urban design and places it in a tropical context. The physical processes behind changing urban climate are clearly illustrated. The book presents a general background to the environmental issues facing tropical building designers – energy consumption and comfort implication of buildings and an overview of microclimate changes brought about by urbanization. New ways of looking at urban design are presented from a climatic design perspective.

The focus of the book is design strategies that can mitigate the negative impacts of tropical urban climate. Three such design goals are: radiant cooling, ventilation and evaporative cooling. Simple design strategies further develop the conceptual ideas, including discussion of a "shadow umbrella" or sun avoidance scheme, and demonstrate their applicability to the equatorial tropics. In order to make the discussions on urban energy use and energy efficiency complete, the book closes the discussions with a series of strategies that facilitate movement between interconnected urban activity patterns in a climatically suitable manner.

Urbanization in tropical regions is beginning to gather momentum and is likely to intensify in the near future. This book fills a crucial gap in the knowledge of climate-sensitive urban design in a tropical context. This comprehensive reference will be welcomed by students and practising architects, as well as by other built environment professionals engaged with the environmental effects of building in worldwide warm and humid climates.

M. Rohinton Emmanuel is a university teacher and researcher from Sri Lanka, with a Ph.D. in Architecture from the University of Michigan, and has also taught in the US and Sweden. His core interests are in studying the environmental changes brought about by urbanization in the warm, humid tropics. For nearly thirteen years he has extensively researched climate changes in cities with the long-term goal of developing urban environmental monitoring indices that facilitate environment-sensitive urban design in the tropics. He has an extensive record of research and publications on building climatology and urban climate changes.

An Urban Approach to Climate-Sensitive Design

Stategies for the tropics

M. Rohinton Emmanuel

Routledge
Taylor & Francis Group
LONDON AND NEW YORK

First published 2005 by Taylor & Francis

Published 2022 by Routledge
4 Park Square, Milton Park, Abingdon, Oxon OX14 4RN
605 Third Avenue, New York, NY 10017

Routledge is an imprint of the Taylor & Francis Group, an informa business

Typeset in Univers by Wearset Ltd, Boldon, Tyne and Wear

Publisher's Note

The publisher has gone to great lengths to ensure the quality of this reprint but points out that some imperfections in the original copies may be apparent.

British Library Cataloguing in Publication Data
A catalogue record for this book is available from the British Library

Library of Congress Cataloging in Publication Data
A catalog record for this book has been requested

ISBN 978-0-415-33409-9 (hbk)
ISBN 978-0-415-33410-5 (pbk)

To Melanie
who makes it possible

Contents

Acknowledgments

This book grew out of the frustration of being unable to find suitable textbooks relevant to the changing climate in tropical cities. Ten years of teaching Architecture in the context of increasing urbanization in the tropics forms the basis of the book. My students, both at the Department of Architecture, University of Moratuwa, as well as at the advanced international training program on "Architecture, Energy and Environment" (AEE) conducted by the Housing Development and Management (HDM) Division of the Lund University, Sweden, helped me decide what is relevant to architects in terms of tropical urban design.

I am particularly indebted to Nirodha Gunadasa and Dhanika Ranasinghe, postgraduate students at the Department of Architecture, University of Moratuwa, for ably drawing the illustrations. I gratefully acknowledge the help of Dr Hans Rosenlund (Course Director – AEE) and Erik Johansson (both from Lund University), in completing the manuscript. Thanks are also due to the reviewers whose ideas and suggestions helped sharpen the focus of the book.

Introduction

CHAPTER 1

1.1 Climate-sensitive design in the tropics: an overview

The awareness that buildings in the tropics should be compatible with climate has existed in ancient civilizations. Codes, design manuals and rules of thumb for the layout of buildings and cities in a climatically suitable manner were known to ancients. The ancient Indian construction manuals such as the "Vastu Sastra" and the "Silpa Sastra" exemplify this knowledge.

Visitors to the tropics readily concede that climate-sensitive design is crucial to human health and well-being in the region. The European explorers documented this need as early as in the eighteenth century. The traditional wisdom, coupled with colonial expansion, led the nineteenth-century colonial buildings to reflect their climatic contexts, though with European overtones.

However, it was only in the 1930s that information and opinions relating to the subject of architecture for the tropics began to be published seriously, France and England being the leading countries where this occurred.

The 1940s saw the broadening of the scope of subjects and deepening of knowledge, in part due to experimentation. One impetus for increased interest and research was the Second World War and the construction difficulties of Americans and Europeans building or guiding building in tropical settings. According to Rivera de Figueroa (1980), the term "tropical architecture" appeared in print for the first time in 1941.

Attempts to avoid difficulties caused by the deterioration of materials in the tropics received considerable attention in the 1940s. The use of wood, iron (particularly with reference to corrosion), glass, aluminum and other lightweight metals, paints, etc., were some of the materials studied in this decade.

The 1950s saw a proliferation of publications relating to the tropics. Specific challenges posed by the tropical climate were the focal concern of this decade. In addition to the *British Colonial Building Notes* series, the US Naval Research Laboratory initiated a series of publications related to construction issues in the tropics. Specific building materials researched in this decade include concrete, bituminous felt and bamboo, in addition to materials already explored in the 1940s.

The most intense search in the 1950s, however, was related to the problem of thermal comfort. Specifically, issues related to comfort in housing began receiving attention at the end of the Second World War and, with the attainment of political independence, this was pursued by many tropical nations. The problem of housing millions of people in these newly independent nations preoccupied the minds of designers. In fact, the word "design" began to be frequently used in publications relating to tropical areas, reflecting a consciousness that architecture could respond to the challenge of climate, not just be a victim of it (cf. de Figueroa, 1980).

The 1960s saw expansion of the building types considered – libraries, hospitals, offices. Specialized educational programs were supported, like the Architectural Association's Tropical Studies Program in London. And it was in the 1960s that books offering comprehensive as well as specialized knowledge of the subject appeared, notably Maxwell Fry and Jane Drew's *Tropical Architecture of the Dry and Humid Zones* (1964). The 1960s was the period of maturation of the subject of design for hot, humid climates.

In terms of publication of textbooks and manuals relating to building types for the tropics, the 1970s were a culmination. Research findings from previous decades led to comprehensive texts such as Koenigsberger *et al.*'s *Manual of Tropical Housing and Building* (1974) and Givoni's *Man, Climate and Architecture* (1976).

The oil embargo of 1973, and the attendant energy crisis, produced a flurry of activities that resulted in the identification of many energy-saving, passive cooling techniques, some of which have applications to the tropics. Through this global event climate-conscious architecture found a strong impetus and, more importantly, research funding. However, a majority of such initiatives came from areas outside the tropical zone (primarily the West), though India also produced some literature in this regard.

Since the oil embargo, interest in "tropical architecture" appears to have waned. The 1980s and the 1990s were generally "barren" of climate-sensitive research and design initiatives. Climatic issues had to compete with qualitative (and therefore "superior") concerns: "meaning" in architecture and architectural tectonics comes readily to mind. Architects and commentators became critical of the narrow predictability of the deterministic climate-conscious architecture of the 1960s and 1970s. Commentators from both within and outside the tropics began to substitute cultural and social factors as shaping forces of architecture. Although it is hard to separate climatic influence from cultural phenomena, such a reaction was understandable in light of the often pedestrian designs that were paraded as climate-sensitive architecture, particularly in the 1960s and 1970s.

Another factor that worked against climate-sensitive design in the 1980s and the 1990s was that even with very high oil prices a case for climate-conscious design on its energy-saving merits seemed less credible as research unearthed the real energy-use patterns of societies. Efficiency improvements in automobiles, home appliances, etc., not only can save more energy than climate-conscious building design but also do so at a lower cost. For example, a US government study shows that substantial energy savings will occur if all windows and openings on a building envelope are sealed and double-glazed metal casements installed (OTA, 1991). However, the energy cost of an average household is too low to make this option economically viable to an average house owner. Suddenly, it seemed that climate-conscious design did not have a substantial rationale behind it.

So, why should designers worry about climate and architecture now?

Ecological sensitivity is one possible and strong reason. If anything would characterize our time it would be the overwhelming concern for the health of the *global* environment. Resource conservation and sustainable consumption will remain dominant concerns, not only for energetic and economic reasons but for environmental, ecological and even spiritual reasons.

The dominant problem of the twenty-first century is not the scarcity of fossil fuels but how "clean" they are. Our collective wisdom at the moment is that wasted energy almost always adds greenhouse gases and causes environmental problems. Any effort to curb the use of energy, therefore, must be welcomed.

A second imperative is to expand our knowledge beyond the climate-deterministic considerations of single buildings. There are vast gaps in our knowledge and understanding of climate-sensitive *urban design*, particularly in the tropics. Climate-sensitive tropical urban design – that is, design of interstices in a climatically sensitive manner – remains largely unexplored. And strategies for implementation and their economic as well as political merits need investigation. We might benefit from new ways of thinking about the subject and a reconsideration of assumptions. What are the boundaries of "urban thermal comfort," for example? In this regard, some crucial developments have already occurred in the fields of environmental engineering and solar engineering, but they need to be translated into architectural and urban design strategies.

Although urban design implications of temperate climates have received some research attention, tropical urban design is rarely mentioned. The relevance of climate-conscious urban design to tropical countries is all the more important now, due to the region's rapid urbanization. As we shall see later, tropical urbanization is going to take mind-boggling proportions in the next fifteen years or so, and no precedence exists for designers to go by. Increasing land prices, fuelled by very high urban population densities, will continue to exert pressure on the urban ecosystem and microclimate changes in urban tropical areas will be large. In an already stressful climate, thermal comfort implications of such changes are very serious. At the same time, it is also a challenge: to create a modern "indigenous" architecture of the region, using its altered climate as a starting point.

What this brief summary points to is that responding to the climatic context has a checkered past in the tropics. However, what has largely been neglected is the climatic suitability of interstitial spaces where a crying need exists for a comprehensive set of design strategies that could, over time, direct the built environment towards a climatically appropriate whole. This need is more pressing in the tropics where the climatic given is very close to the ideals of human comfort, yet the rapid environmental transformation brought about by intense urbanization is making comfort realization an ever-so-elusive goal.

This book is an attempt to cull together the emerging literature on "climate-sensitive urban design" and to place them in a tropical context. Traditional modes of responding to climate are fast becoming obsolete in the urban tropics where higher population densities, changes to air, water and climate qualities and changing lifestyles demand that designers rethink the conventional wisdom of climate-responsive design.

We begin our survey with an overview of microclimate changes brought about by urbanization. Although city-induced environmental changes were intuitively known for centuries, it was only in the last 50 years that empirical evidence to support this intuition has become widely available. In this connection, we place the study of city-induced climate changes in its historical context. We then look at the relatively new evidence for a tropical urban climate anomaly and the known causes for such a phenomenon. A summary of major urban design strategies commonly suggested for the mitigation of the urban climate phenomenon (from other climatic contexts), and a general critique of the same, is also presented.

The goal of climate-sensitive urban design is to achieve human comfort for a majority of urban dwellers. Although the knowledge of thermal comfort in the indoors is well advanced, the study of *urban comfort* lags behind due to the complexities of outdoor radiation balance and the wind regime. The study of urban thermal comfort in the tropics has barely begun. Not that the study of tropical indoor comfort is well settled: on the contrary, even the indoor comfort requirements are confounded by the whole question of thermal adaptation and thermal expectation. But the question of outdoor thermal comfort is more critical in the tropics for two reasons. On the one hand, outdoor living is possible in the tropics for a large part of the year (exemplified by the so-called "usability-factor" of tropical outdoors – Correa, 1989). On the other, the usually porous tropical building skin mandates that outdoor comfort be simultaneously considered together with indoor comfort. Having outlined the climate anomaly in tropical cities, we therefore turn our attention in Chapter 3 to the comfort requirements in tropical urban outdoors where radiation excess, calm winds, high-levels of thermal adaptation and low thermal expectation compound the design problem.

A discussion of the thermal comfort requirements in the urban tropics will quickly illuminate the need for solar access control and the facilitation of air movement in the outdoors. Although solar access rights are well developed in the temperate climatic context, sun avoidance in the urban outdoors is not widely studied. We present some conceptual issues in developing a sun avoidance scheme (called a "shadow umbrella") in Chapter 4, and demonstrate its applicability to the equatorial tropics.

The focus of the book is to point towards design strategies that can mitigate the negative impacts of tropical urban climate anomaly. We present three design goals that attempt to mitigate the tropical urban climate problem in Chapter 5. They are *radiant cooling, ventilation* and *evaporative cooling.* The principles of climatic design at building-level are not necessarily applicable to city-level design decisions. The relative importance of ventilation and radiant cooling is a case in point.

Climate-sensitive urban design in the tropics is not complete without adequate attention being paid to sustainable urban transportation. Sustainable urban transportation hinges on the facilitation of urban transit, not so much on urban transport (i.e. non-mechanized forms of transport). But the problem of movement in the tropics is the ever-so-oppressive outdoor climate, made worse by urbanization. Thus, good-quality-climate-sensitive design in the urban tropics must seek to facilitate urban transit by enlarging the comfort zone in the outdoor. We thus present in the final chapter (Chapter 6) a series of strategies that facilitate movement between interconnected urban activity patterns in a climatically suitable manner.

1.2 Environmental consequences of tropical urbanization

The hot-humid tropics is experiencing both unprecedented and unique urban growth. It is unprecedented because the rate of urban growth in the region has never been equaled. While Europe took almost two hundred years (mid-1700s to early 1900s) to transform from a predominantly rural to a predominantly urban society (Landsberg, 1981), the hot-humid region will do so in less than sixty years (1950–2010) (see, for example, WCED, 1987; Givoni, 1991; Oke *et al.*, 1990/91; Jauregui, 1997 (Figure 1.1). This urbanization is also unique in that the rapid urban population growth was not preceded by industrialization, as was the case in the Western world.

The rapid urbanization has brought in its wake many hitherto unknown changes to humans, other life forms and the physical environment (Detwyler and Marcus, 1972). Changes caused by urbanization on humans include diseases associated with crowding (tuberculosis, pneumonia, respiratory illnesses, measles, common cold, etc.), air pollution-related illnesses and psychological and emotional

1.1 *Rate of urbanization in the tropics.*
Source: Based on data from UN (2002)

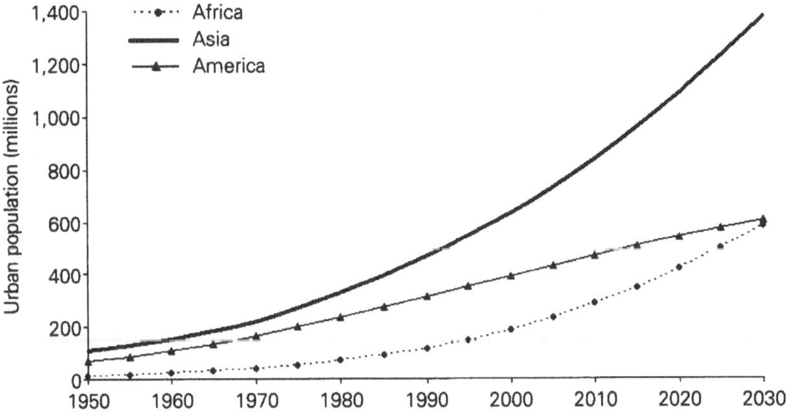

Notes:
Tropical Africa is defined as Western, Middle and Eastern Africa;
Tropical Asia includes South and Southeast Asia;
Tropical America incorporates Central and tropical South America plus the Caribbean;
Tropical Oceania (Micronesia, Melanesia and Polynesia) had less than 2 million urban population in 2000 and is therefore excluded.

disorders (cf. Harrison and Gibson, 1976; Lake *et al.*, 1993). Urban effects on other life forms include physiological changes in urban flora and fauna and their diversity (Sukopp and Werner, 1982), and disease and growth retardation in vegetation (Stulpnagel *et al.*, 1990). Urbanization's effects on the physical environment are apparent in air and water quality (WHO/UNEP, 1992) and the microclimate (see Oke, 1987).

Many of the changes brought about by urbanization are well studied by researchers in diverse fields such as medicine, agriculture and engineering. However, the physical effect of urbanization has not been understood, nor the knowledge base applied, by urban designers, planners, and architects. In particular, the energy and human comfort implications of changes induced by urbanization have received very little attention in urban design and planning. This is especially the case in tropical cities where urbanization is at its peak.

The design and planning professions ought to guide the dynamic environment–city relationship towards more sustainable patterns. Sustainable urban growth is what the World Commission on Environment and Development (WCED, 1987; also known as the Brundtland Report) defined as "growth that meets present needs without compromising the ability of future generations to meet their own needs" (WCED, 1987: 43). However, in the final analysis, responding to the changing urban environment is a quality-of-life (QoL) issue. A positive response to the changing urban environment will be a powerful tool contemporary tropical designers can wield to improve the QoL of their fellow countrymen. While only a minority, albeit a significant minority, of tropical citizens live in cities, their contribution to the creation of their respective national wealth is far beyond their share of population. A deterioration of the QoL of urban dwellers will therefore have a devastating effect on the national economy.

Urban impacts on air quality

In the predominantly non-industrial tropical world, cities are the primary source of air pollution (Schell *et al.*, 1993). Yet urban air pollution (as opposed to atmospheric pollution in general) has not received adequate attention from planners and policy-makers (see Mage *et al.*, 1996). Even the 1992 Rio de Janeiro Conference on Environment and Development, as well as the World Summit on Sustainable Development held in Johannesburg in 2002, did not single out urban air pollution for special attention – although general reference to atmospheric pollution by human activities was made. At the same time, urban air pollution has already become a public health and environmental problem of crisis proportion in many large cities of the world. It is also politically very divisive because "urban air pollution affects every resident, it is seen by every resident, and is caused by every resident" (Mage *et al.*, 1996: 682).

There are three major types of urban air pollutants: greenhouse gases (GHG), aerosols and radiative forcing agents (RFA). These include:

- *Greenhouse gases (GHG)*
  Carbon dioxide (CO_2), carbon monoxide (CO), nitrous oxide (N_2O) and chlorofluorocarbons (CFC).
- *Aerosols*
  Water vapor, sulfur dioxide (SO_2), lead (Pb) and particulates or dust (variously known as total suspended particles – TSP, suspended particulate matter – SPM, or particulate matter – PM. It is customary to specify the diameter of the particulate matter as a subscript. Thus PM_{10} is particulate matter less than 10 microns in aerodynamic diameter.)
- *Radiative forcing agents*
  Ozone (O_3).

A large-scale survey of air quality in 20 of the world's megacities found that particulates (PM_{10}) were the most serious urban air-quality issue. In tropical megacities the levels of SPM frequently exceeded the WHO guidelines by a factor of two (WHO/UNEP, 1992; Mage *et al.*, 1996; Parekh *et al.* 2001). The primary urban SPM emission process is the combustion of fossil fuels for power generation and heating. Additionally, motor vehicles, certain industrial processes and burning of wastes produce SPMs. While soil erosion and other natural processes also emit SPM, human-made (anthropogenic) sources are generally much more toxic (see Beck and Brain, 1984). However, large cities located near deserts or barren lands are susceptible to severe loading of naturally occurring inert crustal SPMs, and their presence in human lungs potentiates the toxicity of the anthropogenic particulates because it increases the residence time of the more toxic SPM (WHO, 1995).

Urban loading of SPMs show clear differences between tropical and temperate cities. In temperate cities SPM levels show no clear seasonal variations while tropical cities experience higher concentration during the dry season and lower levels during the wet season (Panther *et al.*, 1999). During the dry season, SPM concentrations in large tropical cities often exceed the WHO guidelines by a factor of four. However, no statistically significant correlation has been established between rainfall levels and pollutant concentration even in the tropics.

Additionally, cities receive coarser SPMs than rural areas. The urban wind fields, heavily compromised due to the roughness of urban form, not only cause larger particles to be deposited in urban areas but also produce heavier deposition, particularly in city centers (see Erell and Tsoar, 1999) (Table 1.1).

In terms of the aerosols, the main source of anthropogenic emissions of SO_2 is fuel combustion from stationary sources, particularly coal power plants (Stoker and Seager, 1976). Lead, on the other hand, is one of the most common aerosols in cities where leaded petrol is still sold.

Among the RFAs, ozone is the predominant urban air pollutant in tropical megacities. Ozone is not a direct urban pollutant, but is formed by a photochemical reaction of nitrogen oxides and volatile organic compounds (VOC) in the presence of ultraviolet sunlight and moisture. The primary sources of urban VOC include motor vehicles, evaporation of solvents and gasoline, and chemical processing (OTA, 1991). Trees also contribute VOCs to the atmosphere through natural plant processes (so called "biogenic hydrocarbons" – Smith, 1984; McPherson *et al.*, 1998).

The predominant urban GHG is CO_2, its major source being the combustion of fossil fuel. Combustion of fossil fuel contributes approximately 5 billion metric tons of carbon to the worldwide atmosphere annually (McPherson, 1992). However, the level of atmospheric carbon is increasing at the rate of 3 billion metric tons per year; the difference is primarily due to the removal of atmospheric carbon by the oceans (Schneider, 1989). Of this, contribution from urban areas is considerable. In Chicago alone, for example, it is estimated that the annual carbon emission due to automobiles is approximately 1.6 million metric tons (McPherson, 1992).

Motor vehicles are the major anthropogenic source of CO (Stoker and Seager, 1976). Nitrogen oxides (NO_X) arise mainly from the combustion of fossil fuels in urban areas.

Land use development, and conversions too, will affect air quality. The development of vegetated areas could have negative impacts on

Table 1.1 Urban air pollutants and their environmental effects

Pollutant	*Effects*	*Urban sources*
Sulfur oxides	Cause acute and chronic leaf injury; attack wide variety of trees; irritate/impair upper respiratory tracts of animals and people; increased morbidity; destroy paint pigments; erode statues; corrode metals; ruin hosiery; harm textiles; destroy book pages and leather	Fuel combustion, stationary manufacturing processes, power plants
Particulates	Speed up chemical reactions, corrode metals, cause grime on buildings, aggravate lung illnesses, affects cardiovascular system	Automobiles, particularly those with diesel engines
Hydrocarbons	Carcinogenic (i.e. cancer-causing), retard plant growth, cause abnormal leaf and bud development	Automobile, vegetation
Lead	Impairs liver and kidney functions, retards mental development of infants and children, neurological damage	Automobile
Carbon monoxide	Causes headaches, absorbs into blood, reduces oxygen levels, lethal even at small dosage	Automobile
Nitrogen oxides	Visible damage to plant life, irritate eyes and nose, inflammatory and permeability responses; create brown haze, corrode metal	Automobile
Volatile organic compounds	Lead to photo-chemical smog, some evidence of brain damage in laboratory tests on rats. Human effects such as leading to Parkinson's disease are suggested, but not yet proven	Automobile
Ozone	Discolors leaves, reduces crop yields, damages textiles, impedes athletic performance, corrodes a wide range of materials, hastens cracking of rubber, disturbs lungs, irritates eyes, nose and throat, causes headaches and breathing difficulties	Automobile
Small air ions	Positive ions lead to the release of serotonin, which is related to irritable and anxious behavior, heart oppression, migraines and rheumatic pains	Several sources

Sources: Adapted from Stoker and Seager (1976), Smith (1984), WHO, (1987), WHO/UNEP, (1992), and Collier and Hardaker, (1995)

air quality by reducing vegetative and soil absorptive capacity and increasing emissions and ambient air temperatures.

Urban design plays a major role in improving air quality. The avoidance of "urban heat island effect" will greatly enhance pollution dispersal (see Yoshikado and Tsuchida, 1996). This is particularly the case in the tropics where macro-level winds are usually weak. At the local level, proper orientation and design of street canyons (i.e. the space bound by two buildings that face a street) can accelerate pollutant dispersal. On the other hand, differential heating within urban canyons has the potential to inhibit the transport and exchange of air pollutants in urban streets (see Sini *et al.*, 1996).

Additionally, proper management of vegetation in urban areas also has the potential to improve air quality. Vegetation can intercept particulates, absorb various air pollutants and store atmospheric carbon (McPherson, 1992). On the negative side, urban vegetation adds biogenic hydrocarbons (sometimes called biogenic volatile organic compounds – BVOC) and ozone to the city air, the effects of which are only now beginning to be understood (see McPherson *et al.*, 1998; Kuttler and Strassburger, 1999).

The greatest improvement to urban air quality is likely to come from a reduction in automobile use (WHO/UNEP, 1992). As cities expand into megacities, more people will drive more vehicles greater distances and for longer times. It is therefore imperative that motor vehicle inspection programs, phasing out of lead in petrol, promotion of the use of mass transit, a design bias against motor-vehicle oriented development, introduction of road use tax and the introduction of cleaner fuels/vehicles must be encouraged in tropical cities. Although greater improvement in urban air quality will result from technical solutions like clean technologies and fuels, as well as economic incentives such as taxes and rebates, urban designers too need to play an important role in controlling the shape and form of the city which heavily influence the need for motorized transport (see Emmanuel, 1995). The sooner these efforts are implemented, the lower the negative consequences of air quality deterioration (Figure 1.2).

Urban impacts on water quality

Water is a crucial component of the city's support systems. Many of the pollution problems that affect the water system as a whole begin in the city (Hough, 1984: 70).

The main effect of urbanization on water use is threefold: increased usage per capita, quantitative increase of water refuse, and qualitative decrease of the runoff from urban systems. Up to 675 liters of water per person per day are used in cities in the West (the tropical urban average is about 180–200 liters/person/day), as opposed to 18 liters

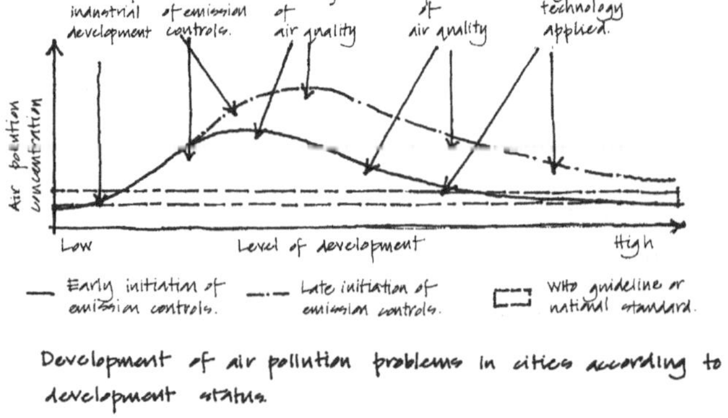

1.2 *Air-quality transition in megacities of the world*

per person per day in rural non-farm communities (Hough, 1984). Almost 85 percent of the rain that falls upon completely built-up urban areas is carried away as runoff by the storm-water drainage system (Lull and Sopper, 1969). This leads to increases in the intensity of runoff (and therefore flashflooding as well as soil erosion and other qualitative damages).

The high-quantity, high-velocity urban storm water also carries with it pollutants (mostly silt and sedimentation materials) that disrupt aquatic life, lead to soil erosion, increase water turbidity and are usually warmer than the rural outflow. High temperature and the presence of many pollutants lead to an increase in biological oxygen demand (BOD), which further threatens aquatic life forms.

The impact of urbanization upon water quality in tropical regions with primate cities is exemplified by the Colombo Metropolitan Region (CMR), Sri Lanka's primate city. The CMR is located largely within the catchment of Kelani river estuary. More than 700,000 inhabitants live within the city limits alone and approximately 50 percent of this population lives in sub-standard settlements habitually prone to flooding. Since the slums of Colombo are largely confined to canal banks and low-lying areas of the city, their vulnerability to flash floods is high, and increasing urbanization increases the likelihood of flash floods in these vulnerable areas. Recent studies have estimated the annual cost of habitual floods in the city of Colombo to be in the region of Rs1.8 billion (US$18 million) (Sri Lanka Land Reclamation and Development Corporation, 1996). This cost is on the rise due to increasing urban population and the increasing hard land cover that seals off the soil in the catchment areas.

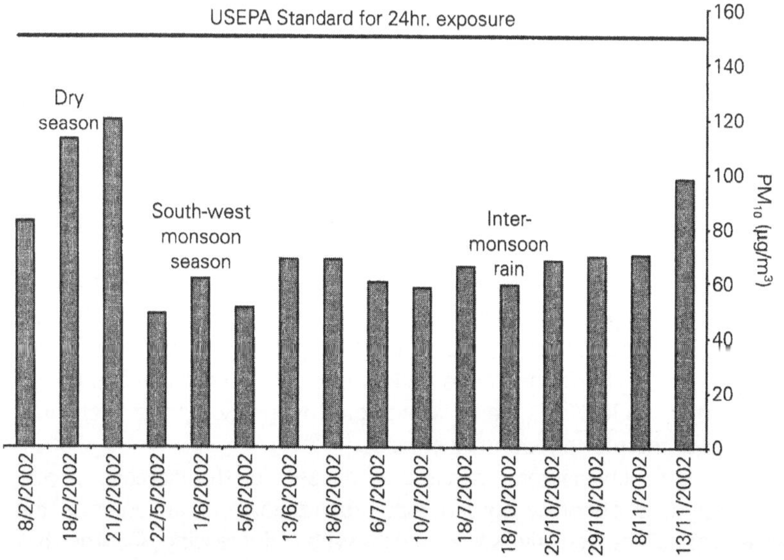

1.3 *Air-quality profile, Colombo Metropolitan Region*

Urban impacts on climate

Of growing concern to urban designers is the rapid change in urban climate associated with urbanization. The phenomenon of city-induced climate change has been known for many centuries. However, the contemporary problem of city-induced climate needs special attention due to the magnitude of the problem. Chapter 2 details the historical evolution of the scientific understanding of the urban climate anomaly and specific studies related to tropical cities, as well as mitigation efforts currently under consideration. Suffice it to say here that urban climate mitigation is crucial to the improvement of thermal comfort, air quality and even water quality in tropical cities, and therefore must be the linchpin of urban physical environment improvement programs.

Critical issues facing climate-sensitive designers in the tropics

In the rapidly urbanizing tropical world, contemporary climate-sensitive designers are forced to operate in an environment where climatic considerations are not the only issue, not even the major issue. We must pay attention to all of them, while focusing our attention on the altering urban climate. At least three issues need to be mentioned in this regard:

1. Building and urban energy use.
2. Fossil-fueled transport.
3. Environmental management.

Building and urban energy use

Patterns of energy use differ among nations, yet certain commonalties exist in their urban energy needs. A comparison of energy-use patterns between a developed country like the US and a developing country like Sri Lanka shows that transportation and activities within buildings consume a considerable share of energy in both cases.

Urban design has an influence on energy needs for transportation and in building energy use. This influence is significant, irrespective of the socio-economic context.

Energy consumption in the developing world has grown tremendously over the past two decades. A World Bank study reports that the developing world, which accounted for 20 percent of the total global consumption in 1970, increased its share to 33 percent in 1988 (Imran and Barnes, 1990). This trend is expected to grow for the first three decades of the present century at least.

Although many factors are cited as causes for the increase in consumption, it is generally agreed that urbanization in the region is the prime cause, particularly so in areas where increasing income has

Table 1.2 Patterns of commercial energy consumption

Activity	*Energy consumption (% of total country needs)*		
	US	*UK*	*Sri Lanka*
Industries	41.2	32.0	9.9
Transportation	21.0	18.0	16.4
Building energy needs	28.0	48.0	67.0
Agriculture/others	7.7	2.0	6.7

Sources: Amerasinghe, (1989), Emmanuel (1995)

resulted in greater demand for transportation and home appliances – especially space-conditioning devices (see de Dear and Fountain, 1994).

One of the main problems with the urban energy consumption lies with the very form of the city (see Hawkes *et al.*, 1987). Urban development is usually piecemeal, left to grow in a haphazard manner, and there are no overall form determinants for most contemporary cities. In the context of a hot-humid climate, energy-efficient urban layout needs to address the following issues (see Emmanuel, 1994):

1. Site and building orientation for avoidance of sun.
2. Site and building orientation for encouraging air movement.
3. Shading device requirements for open spaces.
4. Landscaping requirements for heat gain reduction.

The other important urban energy use is for transportation (Table 1.2). It is also the main source of air pollution as well as of man-made heat input to the urban thermal system. The way urban activities are organized influences the way energy is spent in commuting between them. The greater the diversity of activities at a given place, the less will be the need for mechanized transportation. The solution lies in providing the right mix of activities that will reduce the energy spent on transportation. The goal is not to smooth traffic, rather to reduce it. As Charles Correa says: "ease of transportation is not an end in itself. Journeys always multiply to a point of clogging" (Correa 1989: 73).

Fossil-fueled transport

Table 1.2 indicates that up to a fifth of energy use in developed countries (and somewhat less in the developing world) is consumed by the transportation sector. Energy-efficient urban design cannot ignore the need to encourage energy-efficient modes of transportation. These include walking and bicycles, closely followed by public transportation (see, for example, Illich, 1974; Owens, 1986, 1987; Correa, 1989; Lowe, 1991).

Although there are many shortcomings in the urban transportation strategies employed in tropical cities, the most basic conceptual flaws are twofold: the bias against the poor and the dependence on mechanical power, both of which (as we shall see in Chapter 6) are interrelated. And these flaws are critical from two standpoints of current concern: energy conservation and environmental sustainability.

In terms of energy use the Third World consumption, though small now, is expected to overshadow that of the First World by the year 2020 (Table 1.3). Rapid population growth, increasing affluence (particularly in the urban areas), and structural problems are the causes for this expansion. Most of the energy consumed in the Third World is imported, and the share of oil imports on national income is significant. Most of this commercial energy is consumed by one sector: transportation. Needless to say, urban transportation plays an important part in this conspicuous consumption.

Mechanically biased transportation cannot be successful, particularly in the Third World. The reasons, however, are not just the energy costs indicated above but also the environmental and ecological costs. By "environment" what is meant here is not just the physical environment but also non-physical entities like community, livelihood, freedom, etc. The urban transportation strategies that will work for the resource-starved Third World lie within the paradigm of biased innovations that strengthen the public realm of streets and empower individual freedom of movement at minimal cost, both to oneself and society.

There are three concepts relevant to us: biased innovations,

Table 1.3 Energy dynamics in selected developing countries

Country	*1985*			*1990*		
	1	*2*	*3*	*1*	*2*	*3*
Afghanistan	66.9	42.9	2.3	65.5	54.9	2.3
Bangladesh	51.9	91.5	5.4	43.7	111.2	5.9
Nepal	65.1	73.0	2.6	67.8	128.9	4.0
Pakistan	21.1	45.5	3.4	–	–	–
Sri Lanka	85.9	39.0	8.7	84.3	47.6	10.4
Low-income	*50.7*	*46.8*	*3.6*	*46.0*	*28.5*	*2.3*
Papua New Guinea	91.5	18.8	10.2	87.9	23.0	11.9
Philippines	54.3	34.1	6.8	34.7	29.9	5.5
Thailand	66.3	49.2	10.3	59.3	52.2	10.6
Lower middle-income	*61.5*	*39.0*	*8.3*	*48.3*	*39.6*	*8.1*
China	74.5	34.5	17.8	78.5	53.2	27.5
Fiji	91.1	43.0	6.0	85.4	58.3	6.5
Hong Kong	100.0	13.5	10.4	100.0	16.3	14.3
South Korea	71.6	28.7	10.8	68.9	17.9	9.7
Singapore	100.0	6.3	10.7	100.0	6.2	11.9
High/upper middle-income	*77.1*	*21.2*	*12.5*	*77.9*	*21.4*	*15.2*

Source: Emmanuel, (2001)

Notes: 1 = Share of energy imports in commercial energy consumption (%)
2 = Energy imports as a percentage of export
3 = Energy Imports as a percentage of GNP

empowering the public realm of streets, and freedom of movement at minimal cost. We argue that mechanical transport, both private and mass, is sustainable only at higher capital investment or, worse still, useful only for a small group of people. Therefore, strategies that are most fitting for the Third World are those that emphasize the transportational needs of the poor.

Urban transportation is a function of the overall city form, and lasting structural change can only come from altering the direction of development of the urban form. At the same time, urban design is a piecemeal activity, growing very slowly and at the initiatives of many different people acting in space and time. Design solutions are therefore needed as strategies that can be implemented individually yet direct the growth of urban form towards a sustainable form.

Environmental management

While the urban environmental issues pose significant technical and methodological problems, planners and architects must also address the economics of urban environment management. From the perspective of urban development management, two issues are particularly relevant to the debate on urban development vs. environmental pressures associated with urbanization:

1. Environmental and economic suitability urban construction (i.e. sustainable construction).
2. Urban waste management, particularly waste associated with urban development.

ECONOMIC AND ENVIRONMENTAL COSTS OF URBAN CONSTRUCTION

> Globally, the construction industry is one of the main contributors to the depletion of natural resources and a major cause of unwanted side effects such as air and water pollution, solid waste, deforestation, toxic wastes, health hazards, global warming and other negative consequences. Although the traditional attitude of having unlimited resource and space is still dominant, the awareness of environmental impacts is growing and many movements seeking to address sustainability concerns are gaining momentum. (Augenbroe and Pearce, 2000: 1/17).

The statistics reported by the Ad Hoc Committee on Sustainable Building formed under the ISO/TC59 reveal the contribution of urban construction towards the global environmental pollution and degradation:

1. Approximately 40 percent of the materials produced globally are utilized in buildings and constructions.
2. Approximately 40 percent of all waste globally comes from production and demolition of buildings and constructions.

3. Approximately 40 percent of the total energy consumption globally is related to buildings and constructions.
4. Approximately 40 percent of the CO_2 emission globally is related to buildings and constructions.

These statistics are further reinforced by the Worldwatch Institute, according to which building construction consumes 40 percent of the global use of raw stone, gravel, and sand, as well as 25 percent of the virgin wood. Buildings also account for 40 percent of the energy and 16 percent of the water used annually worldwide (Lippiatt, 1998).

Needless to say, construction related to urban development plays a major part in the industry's contribution to the global environmental problem. Therefore development activities, particularly those in cities, should progressively harmonize with environmental needs. To achieve this, development (particularly urban development) should be viewed in its true perspective: a major consumer of natural resources, a potential polluter of the environment, and yet indispensable.

The true cost of development can only be gauged when all the externalities are factored in. In terms of construction externalities, the economics of environmental management as it relates to urban development must consider at least the following:

1. excessive natural resource consumption, depletion and degradation;
2. waste generation;
3. energy usage.

Conventionally, urban designers and architects select building materials based on cost, aesthetic appearance, availability, ease of installation/construction, durability, and maintenance required during their life period. However, a sustainability perspective would additionally require that urban developers pay attention to the environmental impact from inception to final disposal as well.

However, even the most environmentally conscious building designer or product manufacturer will ultimately weigh the environmental benefits against the economic costs. The environmental performance has to interact with technical and functional performance – and also economic reality. Without this environment sensitive design will fail (see Lippiatt, 1998). This will require further research into all the hidden externalities of urban development in order to bring out the true economic *and* environmental costs of all our developmental actions.

Economics of waste associated with urban development

Urban developmental activities generate enormous amounts of waste. For example, in the USA they contribute approximately 29 percent of overall landfill volumes; in the UK the corresponding figure is over 50 percent and in Australia the range is between 20 and 30 percent

(Rogoff and Williams, 1994; Ferguson *et al.*, 1995). Waste is best defined as any material by-product of human and industrial activity that has no residual value (Serpell and Alarcon, 1998). Although waste is inevitable in all human endeavors, much of the waste that urban development generates has a residual value and is avoidable (Pinto and Agopyan, 1994). This suggests that waste can be reduced, producing benefits for the construction industry and the environment.

Despite being a major generator of avoidable waste, the urban construction industry has been slow to embrace environmentally friendly practices. For decades, landfill has provided a convenient and cost-effective solution to its wasteful habits (Mills *et al.*, 1999). However, resource depletion, global warming, increasing pollution levels, urban population growth, a more educated and outspoken public, and an increasingly legislative business environment are forcing the developmental actors to take a more responsible attitude towards the environment.

Traditionally, research on urban waste management has largely been confined to domestic and industrial waste. Construction waste associated with urban development has received little attention. This is particularly the case in the developing world where urbanization is at its peak and therefore environmental problems associated with construction waste pose major health hazards.

Thus climate-sensitive architecture in the tropics ought to be cognizant of the following contextual issues: changing microclimate, increasing urbanization and associated changes to human quality of life, reducing resource base, increased energy use – particularly for transport, growing urban population density and the growing problem of waste disposal (particularly construction waste). Conventional understanding of these issues is constantly being questioned.

References

Amerasinghe, D. (1989). "Impact of energy price changes on manufacturing industries," in E.N. Wijemanne (ed.), *Energy Pricing Options in Sri Lanka*. New Delhi: International Labor Organization/Asian Regional Team for Employment Promotion, pp 18–53.

Augenbroe, G.L.M. and A.R. Pearce (2000). "Sustainable construction in the USA: perspectives to the year 2010," in P. Brandon, P. Lombardi, and S. Perera (eds), *Cities and Sustainability: Sustaining Our Cultural Heritage*, Proceeding of the Millennium Conference, University of Moratuwa, Sri Lanka, pp. 1/17–1/25.

Beck, B.D. and J.D. Brain (1984). "Prediction of the pulmonary toxicity of respirable combustion products from residential wood and coal stoves," in *Residential Wood and Coal Combustion*, Louisville, Ky.: Air Pollution control Association, pp. 264–280.

Collier, C.G. and P.J. Hardaker (1995). "Weather, air quality and health," *Meteorological Applications*, 2(4): 313–322.

Correa, C. (1989). *The New Landscape: Urbanization in the Third World*, London: Butterworth Architecture.

de Dear, R. and M.E. Fountain (1994). "Field experiments on occupant comfort and office thermal environments in a hot-humid climate," *ASHRAE Transactions*, 100(2): 457–475.

de Figueroa, R. (1980). *Architecture for the Tropics: A Bibliographical Synthesis (from the beginnings to 1972)*, Rio Piedras, Puerto Rico: Editorial Universitaria, University of Puerto Rico.

Detwyler, T.R. and M.G. Marcus (1972). *Urbanization and Environment: The Physical Geography of the City*, Belmont, Calif.: Duxbury Press.

Emmanuel, R. (1994). "Recovering the commons: a new approach to climate-conscious urban design in the equatorial tropics," Paper presented at the 12th Annual ACSA Technology Conference, in Jong-Jin Kim and Colin Clipson (eds), *Design and Technological Innovation for the Environment*, Washington, D.C.: Association of Collegiate Schools of Architecture (ACSA), pp. 124–128.

Emmanuel, R. (1995). "Energy-conscious urban design guidelines for warm-humid cities: strategies for Colombo, Sri Lanka," *Journal of Architectural and Planning Research*, 12(1): 58–75.

Emmanuel, R. (2001). "Strategies for equity and sustainability in urban transportation in Third-World mega-cities," *Built Environment – Sri Lanka*, 2(1): 20–34.

Erell, E. and H. Tsoar (1999). "Spatial variations in the aeolian deposition of dust – the effect of a city: a case study in Be'er-Sheva, Israel," *Atmospheric Environment*, 33: 4049–4055.

Ferguson, J., N. Kermode, C.L. Nash, W.A.J. Sketch, and R.P. Huxford (1995). *Managing and Minimizing Construction Waste: A Practical Guide*, London: Institution of Civil Engineers.

Fry, M. and J. Drew (1964). *Tropical Architecture of the Dry and Humid Zones*, London: Architectural Press.

Givoni, B. (1976). *Man, Climate and Architecture*, London: Applied Science Publishers.

Givoni, B. (1991). "Impact of planted areas on urban environmental quality: a review," *Atmospheric Environment B*, 25(3): 289–299.

Harrison, G.A. and J. B. Gibson (eds) (1976). *Man in Urban Environments*, Oxford: Oxford University Press.

Hawkes, D. *et al.* (eds) (1987). *Energy and Urban Built Form*, Papers presented at the International Seminar on Urban Built Form and Energy Analysis, June 26–27, 1986, Darwin College, Cambridge, UK: Centre for Configurational Studies, Open University/Martin Centre for Architecture Studies.

Hough, M. (1984). *City Form and Natural Processes*, New York: Van Nostrand Reinhold.

Illich, I. (1974). *Energy and Equity*, London: Calder and Boyars.

Imran, M. and P. Barnes (1990). *Energy Demand in the Developing Countries: Prospects for the Future*, World Bank Staff Commodity Working Paper No. 23, Washington, D.C.: The World Bank.

Jauregui, E. (1997). "The last M's for 40th anniversary issue: aspects of urban human biometeorology," *International Journal of Biometeorology*, 40: 58–61.

Koenigsberger, O.H., T.G. Ingersoll, A. Mayhew, and S.V. Szokolay (1974). *Manual of Tropical Housing and Building, Part I*: Climatic Design, London: Longman.

Kuttler, W. and A. Strassburger (1999). "Air quality measurements in urban green areas – a case study," *Atmospheric Environment*, 33: 4101–4108.

Lake, J.V., G.R. Bock, and K. Ackrill (eds) (1993). *Environmental Change and Human Health*, Chichester, UK: Wiley-Interscience.

Landsberg, H.E. (1981). *The Urban Climate*, New York: Academic Press.

Lippiatt, B.C. (1998). *Balancing Environmental and Economic Performance*, Gaithersburg, Va: National Institute of Standards and Technology.

Lowe, M. (1991). "Bicycle, vehicle for a small planet," *Ekistics*, 58(348/349): 179–194.

Lull, H.W. and W.E. Sopper (1969). *Hydrologic Effects from Urbanization of Forested Watersheds in the Northeast*, USDA Forest Service Technical Report NE-146, Upper Darby, Pa.: USDA Forest Service.

McPherson, E.G. (1992). *Chicago's Evolving Urban Forest*, Chicago, Ill.: USDA Forest Service, Northeastern Forest Experiment Station.

McPherson, E.G., K.I. Scott, and J.R. Simpson (1998). "Estimating cost effectiveness of residential yard trees for improving air quality in Sacramento, California, using existing models," *Atmospheric Environment*, 32(1): 75–84.

Mage, D. *et al.* (1996). "Urban air pollution in megacities of the world," *Atmospheric Environment*, 30(5): 681–686.

March, L. and L. Martin (1972). *Urban Space and Structure*, Cambridge: Cambridge University Press.

Mills, T.H., E. Showalter, and D. Jarman (1999). "A cost effective waste management plan," *Cost Engineering*, 41(3): 35–43.

Office of Technology Assessment (OTA) (1991). *Energy Technology Choices: Shaping our Future*, Washington, D.C.: OTA.

Oke, T.R. (1987). *Boundary Layer Climates*, London: Methuen.

Oke, T.R., R. Taesler, and L.E. Olsson (1990/91). "The TRopical Urban Climate Experiment (TRUCE)," *Energy and Buildings*, 15/16: 67–73.

Owens, S. (1986). *Energy, Planning and Urban Built Form*, London: Pion Ltd.

Owens, S. (1987). "The urban future: does energy really matter?" in D. Hawkes *et al.* (eds), *op. cit.*, pp. 169–186.

Panther, B.C., M.A. Hooper, and N.J. Tapper (1999). "A comparison of air particulate matter and associated polycyclic aromatic hydrocarbons in some tropical and temperate urban environments," *Atmospheric Environment*, 33: 4087–4099.

Parekh, P.P. *et al.* (2001). "Ambient air quality of two metropolitan cities of Pakistan and its health implications," *Atmospheric Environment*, 35: 5971–5978.

Pinto, T. and V. Agopyan (1994). "Construction wastes as raw materials for low-cost construction products," in C.J. Kibert, (ed.), *Sustainable Construction*, Gainsville, Fla.: Center for Construction and Environment, pp. 335–342.

Rogoff, M.J. and J.F. Williams (1994). *Approaches to Implementing Solid Waste Recycling Facilities*, Park Ridge, N.J.: Noyes.

Schell, L.M., M.T. Smith, and A. Billsborough (1993). "Human biological approaches to the study of third world urbanism," in L.M. Schell, M.T. Smith, and A. Billsborough (eds), *Urban Ecology and Health in the Third World*, Cambridge: Cambridge University Press, pp. 1–9.

Schneider, S.H. (1989). "The changing climate," *Scientific American*, 261: 70–79.

Serpell, A. and L.F. Alarcon (1998). "Construction process improvement methodology for construction projects," *International Journal of Project Management*, 16(4): 215–221.

Sini, J.-F., S. Anquetin, and P.G. Mestayer (1996). "Pollutant dispersion and thermal effects in urban street canyons," *Atmospheric Environment*, 30(15): 2659–2677.

Smith, W.H. (1984). "Pollutant uptake by plants," in M. Treshow (ed.), *Air Pollution and Plant Life*, New York: John Wiley and Sons, pp. 417–450.

Stoker, H.S. and S.L. Seager, (1976). *Environmental Chemistry: Air and Water Pollution* (2nd. edn), Glenview, Ill.: Scott, Foresman and Co.

Stulpnagel, A V., M. Horbert, and H. Sukopp (1990). "The importance of vegetation for urban climate," in H. Sukopp, S. Hejny, and I. Kowarik (eds), *Urban Ecology: Plant and Plant Communities in Urban Environments*, The Hague: SPB Academic Publishing Co., pp. 175–193.

Sukopp, H. and P. Werner (1982). *Nature in Cities*, Nature and Environment Series No. 28, Strasbourg: Council of Europe Publication.

United Nations (UN) (2002). *World Urbanization Prospects – 2001 Revision*, New York: United Nations, Population Division, Department of Economic and Social Affairs.

WHO (1987). *Air Quality Guidelines for Europe*, World Health Organization Regional Publications, European Series No. 23, Copenhagen: WHO Regional Office.

WHO (1995). *Interpretation of WHO Air Quality Guidelines: Report of an Expert Working Group*, WHO/EOS/95, Geneva: World Health Organization.

WHO/UNEP (1992). *Urban Air Pollution in Megacities of the World*, World Health Organization, United Nations Environmental Program, Oxford: Blackwell.

World Commission on Environment and Development (WCED) (1987). *Our Common Future*, Oxford: Oxford University Press.

Yoshikado, H. and M. Tsuchida (1996). "High levels of winter air pollution under the influence of the urban heat island along the shore of Tokyo Bay," *Journal of Applied Meteorology*, 35(10): 1804–1813.

Urbanization and climate CHAPTER 2

2.1 Historical background

The phenomenon of city-induced environmental change has been known for many centuries. The ancient Indian architectural manual "Silpa Sastra" (translated by Acharya, 1979) laid out rules for the siting of villages, towns and forts based on prevailing wind directions and solar orientation. Building and street layout, and massing, were prescribed so as to enhance street-level shade, air-pollution dispersal and storm-water drainage. Vitruvius' (75–25 BC) Book I of *The Ten Books on Architecture* (trans. Morgan, 1960) also considered city layout in relation to local environmental conditions.

The semi-scientific awareness of the urban climate anomaly dates back to at least the mid-eighteenth century. For example, in the 1750s Ezra Stiles noted that the air of towns was warmer than that of the countryside (Meyer, 1991). Charles Caldwell (1798) opined that Philadelphia possessed "a summer temperature superior by three to four degrees, to that of the country." William Currie (1792) attributed the narrowness and irregular form of houses and the crowded building habits as causes for New York's high temperatures in the summer. Noah Webster (1799) cautioned meteorologists of his day about the "error of determining temperature ... by observations made in a large city."

The early interest in city-induced climate changes grew out of a concern for human health. For the sake of street-level thermal comfort, Palladio recommended that in cold places the street should be "ample and broad" while in hot places they should be "narrow with high houses" (see Bosselmann *et al.*, 1995). Caldwell (1833: 367) suggested "city houses [be made of] thick walls, small windows and spacious apartments surrounded by trees of sufficient elevation to shade them." For the hot, humid city of New Orleans, Caldwell (1836: 634) opined narrow streets and overhead balconies would be appropriate.

However, the experiential awareness of city-induced climate changes and intuitive knowledge of mitigation strategies were not systematically studied until the development of meteorological instruments such as the thermometer, the barometer, and the anemometer (see Oke, 1991: 61). The widespread availability of meteorological instruments in the mid-nineteenth century heralded what could now be termed as the "scientific" investigation of urban microclimate modifications.

Luke Howard (1833) is credited to have carried out the first scientific study of inadvertent urban climate modifications (Landsberg, 1981). He compared the temperature records of a city weather station (London) with that of a (then) rural station (Kew Gardens) and found that the city station was warmer. The warmth of cities in contrast to their rural surroundings is now called the "urban heat island" (UHI), a term – according to Landsberg (1981) – probably coined by Gordon Manley (1958).

2.2 Urban heat island: the phenomenon

A heat island is best visualized as a dome of stagnant warm air over the heavily built-up areas of cities (see Figure 2.1). UHIs have been observed practically in all parts of the world except in extreme cold climates. Contrary to the conventional wisdom, heat islands are intense at night, occurring a few hours after the sunset. This has far-reaching implications for urban design in the tropics where night-time thermal stress is comparatively low (Table 2.1).

A clear indication that a city has a UHI is the rapid rise in minimum (night-time) temperature and a slow rise in maximum (daytime) temperature: thus, the daily (i.e. diurnal) variation between the maximum and the minimum temperatures will diminish over time (Landsberg, 1981: 87).

Causes of UHI

Research in the recent past has unearthed a series of causal relationships between a wide range of urban factors and climate. These range from urban geometry to work-week patterns, from anthropogenic heat to thermal characteristics of urban surfaces, and from obstruction to wind flow to lack of vegetation. However, the widely prevalent view among urban climatologists is that at neighborhood and smaller scales

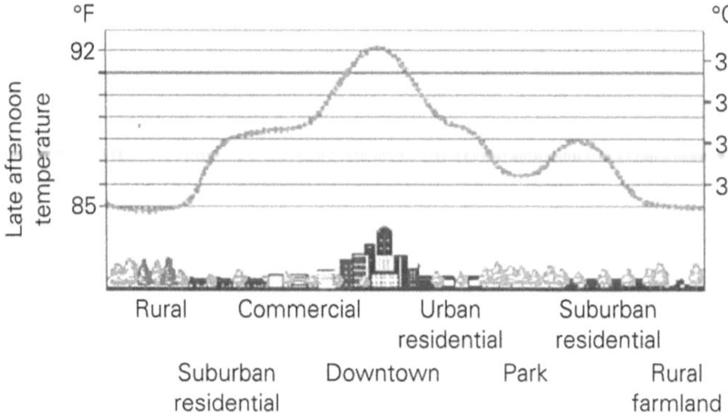

2.1 *Graphical representation of a typical UHI.* Source: Akbari et al. (1992)

Table 2.1 Effect of urbanization on climatic parameters

Climatic parameter	*Effect of urbanization*
Temperature	Rise in daily minimum temperature: Some change in maximum temperature.
Humidity	Reduction in daytime humidity, but increase in night-time values.
Precipitation	Larger increases in summer (up to 21 percent) and smaller increases in winter (5–8 percent). In the tropics, the increase is attributed more to air pollution than heat emission.
Wind	Increases in the number of calm periods observed. Up to 20 percent reduction in wind speeds are known. The effect is greater upon weaker winds.
Solar radiation	Though incoming radiation values are not changed, the apparent values are high due to the containment of reflected radiation by the heat dome.

Sources: Khemani *et al.* (1973), Chandler (1976), Landsberg (1981), Oke (1987), Emmanuel (1997c)

urban geometry leads the list of possible causes for the heat-island phenomenon.

The urban geometry of a city is characterized by a repetitive element called the urban canyon. Urban canyon is defined as the three-dimensional space bounded by a street and the buildings that abut the street. Urban canyons restrict the view of the sky dome (characterized by the sky view factor SVF), cause multiple reflection of solar radiation, and generally restrict the free movement of air (see Figure 2.2). For long urban canyons it is customary to specify the geometry by its height of building/width of street (H:W) ratio, sometimes known as the aspect ratio.

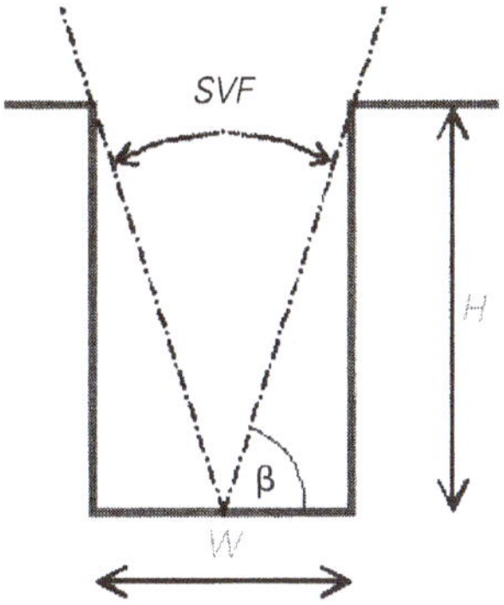

Canyon geometry for long canyons = height of buildings (H)/width of street (W)

Sky view factor (SVF) = fraction of sky visible at middle of street

For infinitely long canyons, SVF = Cos β

2.2 *Canyon geometry and SVF*

Urban geometry and heat island effects

Traditionally, thermal properties of building materials were thought to be the primary cause for the urban climate anomaly. However, evidence indicating the importance of urban geometry to the creation and the magnitude of the UHI phenomenon began to be accumulated in the 1970s. Terjung and Louie (1973) were among the first to suggest that the urban/rural daytime temperature anomaly is largely attributable to the aspect of vertical surfaces to the sun. Goward (1981) showed that the daytime urban heat excess cannot be attributed solely to the higher thermal inertia of urban surfaces relative to rural areas.

In Lawrence, Kansas (a small town 25 miles east of the state capital, Topeka), Henry and Dicks (1987) found that the temperature difference between a highly developed part of the town and a representative rural area to be actually 1.9°C (3.4°F) higher in the rural case. This anomaly was attributed to the almost city-like street geometry of that particular rural setting.

Comparing the UHI effect of thermal properties and urban geometry, Oke (1981) found no significant correlation between types of man-made surfaces and heat-island intensity. The only significant difference observed was between man-made and natural surfaces. Thus, concrete surfaces and asphalt paving, brick walls and cinder blocks, all contribute more or less equally to the problem of urban heat build-up, while tree-covered areas show remarkable reduction in heat build-up.

Noting that the urban geometry factors have not received as much consideration in explaining the spatial–temporal distribution of UHIs as surface materials, Todhunter (1990a) took the position that at the micro-scale, explicit considerations of urban geometry were more important. At the meso-scale, however, he argued that both the geometry and surface thermal characteristics played an equal role.

In what could be termed as a pioneering synthesis of the knowledge base concerning the urban geometry effects, Arnfield (1990) compared the canyon geometry effects with those by the thermal properties of the urban fabric, and concluded thus:

1. Canyon geometry alone is capable of giving rise to variations in nocturnal cooling which can generate a UHI.
2. Differences among the cooling rates exhibited by different canyon geometries tend to be enhanced by [differences] among the thermal properties of the urban fabric . . . If the density of wall materials is correlated with the aspect ratio (heavier construction materials in downtown areas and lighter ones in the suburbs), the differences in the thermal responses of different canyons are enhanced.
3. Wall thickness acts in the opposite direction to (2) above; thicker walls in downtown areas will tend to reduce the differences in thermal response. However, this . . . is unlikely to eliminate the pattern determined by the geometry and materials alone.
4. Cloud amount and type play major roles in . . . decreasing the differences among different canyon forms (pp. 236–237)

He also indicated that if heat-absorbing materials are used in wrongly oriented urban canyons, the effect of the heat island will be more pronounced.

Other causes of UHI

The microclimate effects of urban geometry and surface thermal properties are further augmented by human-made (anthropogenic) waste heat input and alterations to the wind and water balance in cities (Figure 2.3):

2.3 *Causes for UHIs*

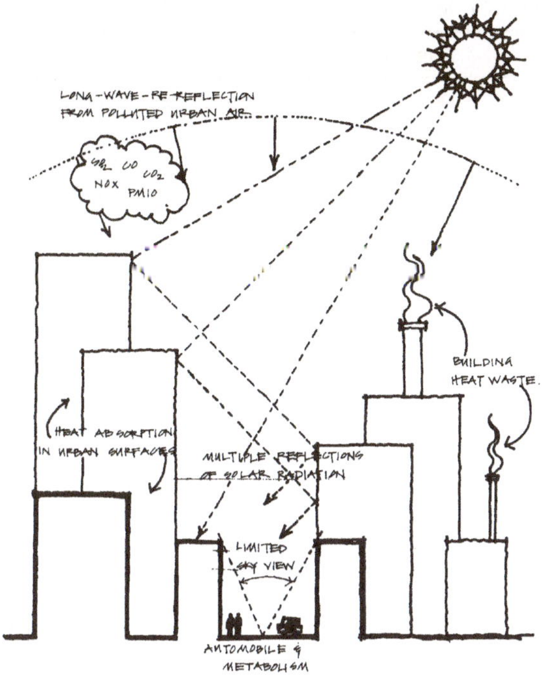

1. Anthropogenic heat: heat waste from combustion and metabolism.
2. Urban "greenhouse" effect: increased incoming long-wave radiation from polluted urban atmosphere.
3. Evapotranspiration loss: reduced green spaces in cities leads to more sensible than latent heat transfer;
4. Wind shelter: reduced ability of wind to carry heat (turbulent transfer) either as sensible or latent heat (see Todhunter, 1990a: 131–132; Oke *et al.*, 1991: 339–340; Saitoh and Hoshi, 1993).

The magnitude of urban climate modifications also depends on the synoptic weather conditions (Todhunter, 1990b; Oke *et al.*, 1991; Kidder and Essenwanger, 1995) and the time of day. The maximum urban–rural air temperature differences occur soon after sunset under calm, clear skies (Landsberg, 1981; Oke, 1987). Temperature differences tend to dissipate under windy or overcast conditions. However, surface temperature differences peak soon after solar noon (see Outcalt, 1972b; Pease *et al.*, 1976).

The physics of urban climate anomaly is presented in greater detail later. Although many different approaches have been taken in explaining urban heat islands, the energy-based approach presented on pp. 28–29 not only explains the causes but also points to urban design solutions. Suffice it is to say here that the partitioning of more heat into sensible rather than latent form is the primary energetic effect of urbanization (see also Table 2.2).

Table 2.2 Microclimatic effects of city parameters

Phenomenon	*Possible effect*
Population Size	Urban-rural temperature difference is proportionate to population size. The effect is pronounced as population increases beyond one million. However, the relationship is much more complicated in tropical cities.
Topography	Air drainage caused by elevation differences and secondary wind patterns created by topography tend to reduce the heat-island effect.
Rivers and other waterbodies	UHI dissipates at or near waterbodies, even if the waterbody is in the middle of a heat island.
Wind speed	UHI intensity is inversely proportional to macro-level wind speeds. However, urban wind flows are usually weak at macro-level.
Anthropogenic heat	The presence of large numbers of heating equipment and air conditioners can aggravate the heat-island effect, though the effect pales in comparison to solar energy trapped by buildings.
Water runoff	UHI effect is more pronounced if more rainwater is allowed to drain away quickly from cities. Availability of water helps partition more heat by evaporation.
Vegetative cover	The presence of sufficient vegetative cover not only ensures more evaporation but also less radiation received at the earth's surface (due to photo-synthesis)

Sources: Chandler (1976), Landsberg (1981), Oke (1982)

Classification of urban heat-island literature

The scientific investigations of urban microclimate modifications fall into three categories: descriptive studies, diagnostic studies, and application studies. Oke (1991) used the following chronology to distinguish the first two. Prior to 1930: simple climatographies of individual weather elements; from 1930 to 1965: description was extended to complex spatial and temporal variability and linkage of climate effects to weather and urban structure; post-1965: emphasis shifted to seek the physical basis of urban climate modifications and to construct models to simulate them (diagnostic studies). In the recent past, the emphasis has shifted to mitigation aspects, particularly in light of the concerns highlighted by the Kyoto Protocol, the Rio de Janeiro Conference on Sustainable Development, as well as by the World Summit on Sustainable Development (WSSD) held in Johannesburg in 2001.

The oldest form of urban climate research – descriptive studies – employed either the city–rural comparison method or the automobile traverse. The former is easily accomplished by a simple comparison of historic climatic records from a city weather station and a nearby rural station where "contamination" due to urbanization is thought to be minimal. City traverse involves mobile measurements of weather parameters (usually air temperature and relative humidity) several times across the urban areas and comparing the readings against a reference station established for the purpose. In recent years, these approaches have been supplemented by remote sensing techniques.

Case studies of individual cities are an early form of descriptive studies (Schmidt, 1929 – Vienna; Sundborg, 1950 – Uppsala, Sweden; Nieuwolt, 1966 – Singapore; Sani, 1973 – Kuala Lumpur, Malaysia, etc.). In more recent times, descriptive studies have taken the form of surveys of urban climate in many cities of a region (for example: Switzerland – Wanner and Hertig, 1984; Tama river basin, Japan – Yamashita *et al.*, 1986; Southwest US – Balling and Idso, 1989; several tropical cities – Tso, 1996).

The descriptive research established the parameters of study and defined the nature of urban microclimate modifications. The results of these studies were presented in some of the ground-breaking conferences in the field (for example, the 1st International Conference on Building and Urban Climatology, Belgium, 1968 [WMO, 1970]). Exhaustive bibliographies also appeared during this period (for example, Landsberg, 1956; Chandler, 1965; Peterson, 1969; Oke 1974, 1979).

The diagnosis of causes of urban climate alterations was the prime research objective during the late 1960s to the early 1980s. The search for causes was greatly facilitated by the distinction Oke (1976) drew between the urban canopy layer (UCL) and the urban boundary layer (UBL). UCL is the space between the earth's surface and the average height of roofs. The atmosphere immediately above the UCL is considered to be the UBL (see p. 28 for a brief description).

The diagnostic studies were typically experimental in design and were carried out either in natural (Oke, 1981) or artificial (simulation) settings (Vukovich *et al.*, 1976; El-Sioufi, 1987; Oke *et al.*, 1991; Sailor, 1995). This period also saw the publication of more specialized bibliographies (Oke, 1974: historical reviews; Oke 1982, 1988a: urban energetics; Jauregui, 1993, 1996: tropical urban climatology) and conferences (for example, 1st Urban Climate Conference with Special Regard to Tropical Areas, Mexico City, 1984). This tradition has been carried forward to this day by the International Association of Urban Climate (IAUC): 1st International Conference of Urban Climate, Planning and Building, Kyoto, Japan, 1989, 2nd Conference in Dhaka, 1993, 3rd in Essen, 1996, and the 4th in Sydney 1999, now in collaboration with other like-minded organizations (for example, the International Congress of Biometeorology [de Dear and Potter, 1999]).

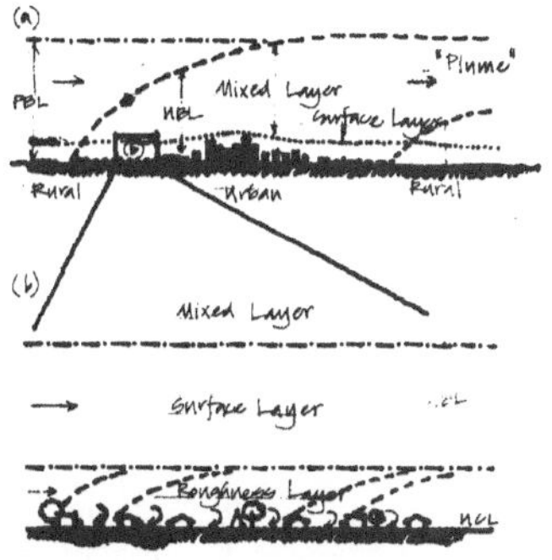

2.4 *Distinction between the UBL and UCL.*
Source: Oke (1987)

The diagnostic efforts resulted in the creation of several UHI simulation models. These include, URBAN3 model (Myrup, 1969, modified by Outcalt, 1972a); URBD model (Terjung and Louie, 1974); URBMET model (Bornstein, 1975) and the HFLUX model (Carlson *et al.*, 1981).

The focus of the more recent UHI models has changed towards facilitation of better UHI measurement as well the evaluation of mitigation options. These include Saitoh *et al.* (1996), using a multilateral element method; the Cluster Thermal Time Constant (CTTC) Model (Swaid and Hoffman, 1990a, 1990b; Swaid *et al.*, 1993); the Green CTTC Model (Shashua-Bar and Hoffman, 2002a, 2002b); and the more numerical models such as Mills (1993, 1997) and Baik and Chun (1997).

The energetic processes of the urban heat island

The understanding of urban energy balance is greatly enhanced by Oke's (1976) conception of the urban atmosphere as a system composed of two distinct layers: the urban canopy layer (UCL) and the urban boundary layer (UBL). The UBL is the overall atmospheric system that extends for many miles above cities (see Figure 2.4). The characteristics of the UBL are partially determined by the city below (Oke, 1982: 2). The UCL, on the other hand, is that layer of atmosphere where most life occurs: from ground up to the mean height of roofs. Understandably, climatic effects of urbanization are strongly felt in the UCL.

We know from the laws of thermodynamics that the energy input into a system must either be released from the system (output) or stored in the system to be released at a later time.

The primary energy input into an urban system is the radiation from the sun. Human activities, too, add some heat into an urban system. We call it the "anthropogenic heat." Solar radiation could either reach the earth's surface as direct (short-wave) radiation or heat up the atmosphere first and then be directed towards the earth's surface (usually long-wave). The energy that reaches the earth's surface may be immediately reflected back (short-wave) or stored and then lost to the atmosphere (long-wave). Additionally, energy may be lost due to convective flow (called the turbulent flow). In this category we include energy that heats up the urban system and is then convected out of it (sensible heat) as well as the energy that heats the water in the system and then is convected out as evaporation. Water that transpires from vegetation should also be included in the latter category. Together, these two convective processes (evaporation and transpiration) use latent heat transfer and are commonly called the evapotranspiration.

Oke (1987) provides a comprehensive treatment of the urban energy flows. According to Oke (1987), urban areas significantly alter both the sources and the sinks of the heat flow mechanisms. Two heat sources (net all wave radiation and anthropogenic heat) and three heat sinks (sensible heat loss, latent heat loss, and storage) are important players in urban thermodynamics, both at UCL and the canyon level. In terms of heat sources, a beneficial arrangement of less short-wave radiation intake and more long-wave radiation intake seems to occur in cities. It also appears that cities lose more long-wave output but less short-wave reflected radiation. As such, the overall effect of sun and atmosphere as heat sources appears to remain unchanged, at least for mid-latitude cities (Grimmond and Oke, 1995). The case of tropical cities is not clear (Jauregui *et al.*, 1992).

Industrial and manufacturing activity, automobile heat rejection, and building energy wastage appear to be significant heat sources for urban areas. However, there are wide temporal and spatial variations in the intensities of such anthropogenic heat sources. Seasonally, heating seasons emit more building energy waste than the cooling seasons. Diurnally, night-time anthropogenic emission seems more important than the daytime amounts. Spatially, downtown areas contribute more per unit area than the suburbs and residential zones of cities. Overall, however, the role of a spatially and temporally averaged anthropogenic heat input seems relatively minor. Although data does not exist for tropical cities, it is not likely to be different as the economic viability and climatic need for greater energy wastage do not exist in that region.

On the sink side of the equation, heat storage in the urban fabric appears to be the primary flow mechanism at the urban canopy level. At the boundary layer level, however, heat storage plays only a modest part in trapping urban excess heat.

It is in the relative partitioning between sensible and latent means of heat flow that urban areas appear to exert the most influence. At the canyon level, canyon geometry and surface thermal properties dictate the relative proportions of sensible and latent heat loss. At the boundary level, moisture availability and, by inference, the presence/absence of vegetation appear to do the honors.

At meso-scale or UBL level, urban geometry-induced heat alterations are minor due to the canceling effects of canyons of various sizes, geometries, orientations and thermal properties found all over the city. At the canyon level, however, explicit consideration of the urban geometry induced heat flow processes is necessary (Todhunter, 1990a). Furthermore, the thermal influence of canyon geometry appears to recede with the distance from the city center (Barring *et al.*, 1985). The surface temperatures, however, remain strongly related to the canyon geometry, irrespective of its location in relation to the city center.

We may thus conclude, with Lewis and Carlson (1989), that "the surface energy responses [at the canyon level] may define themselves through some mix of canyon geometry, thermal inertia [thermal properties of surfaces] and moisture availability" (Lewis and Carlson, 1989: 203).

2.3 Tropical UHI studies

Although the urban climate of tropical cities has not been extensively studied, the few studies that exist (Table 2.3) suggest that the micro-climatic changes due to urbanization in the region are substantial and the anatomy of tropical heat islands bears a close resemblance to that of temperate ones (Oke, 1982: 6).

Table 2.3 Recent tropical UHI studies

Author/s	*City*	*Parameter/s studied*	*Major finding*
Nichol (1996a, 1996b)	Singapore	Remotely sensed surface temperature	Due to high solar azimuth, horizontal surface temperatures are more representative of urban air temperatures in the tropics. Tropical cities do not have a single UHI; rather a collection of small UHIs separated by cooler areas.
Jauregui and Romales (1996)	Mexico City	Convective precipitation	Wet season rainfall, as well as the frequency of intense rainfall (> 20 mm/hr), appears to have increased over the city. The latter is related to daytime UHI.
Jauregui and Tejeda (1997)	Mexico City	Specific humidity	City is drier during the day and wetter during the night than rural areas. The city–rural differences also depend on the season (smaller during dry season and larger during wet season).
Jauregui (1997)	Mexico City	Air temperature	Nocturnal heat island was more frequent (75 percent of the time) than daytime heat island (25 percent). Daytime heat island may have been caused by differences in evaporative cooling from wet surfaces during the wet season.
Jauregui *et al.* (1997)	Mexico City	Air temperature and relative humidity	The heat-island effect reduces the "cold" nights to "cool" and "cool" nights to "comfortable" bioclimate (as measured by effective temperature – ET).
Barr-Kumara-kulasinghe (1997)	16 cities in South India and Sri Lanka	Air temperature	Negligible temperature trends were seen in all but one city (Colombo, Sri Lanka).
Deosthali (1999)	Pune, India	Wet and dry bulb temperature	Rising trends in annual and monthly thermal comfort (THI), particularly during the day. The presence of a "moisture island" detected.
Oke *et al.* (1999)	Mexico City	Net radiation, sensible and latent heat fluxes	During daytime, the heat uptake by buildings is so large that convective heating is severely suppressed in the central city with massive stone walled buildings. The heat release at night is equal to or larger than net radiation.
Wienert and Kuttler (2001)	Several cities	Air temperature differences between city and rural areas	UHI magnitude is linked to latitude (low latitudes have smaller UHI), but this correlation is largely explained by differences in anthropogenic heat and radiation balance.
Emmanuel (2003)	Colombo	Air temperature and relative humidity	Thermal comfort patterns (THI) are strongly correlated to hard land cover changes, particularly in the suburban areas.

One of the earliest tropical urban microclimatic studies was conducted by Nieuwolt (1966) in Singapore. Taking air temperature and relative humidity measurements at various points in the city and suburban areas, and comparing them with the readings from the Singapore Airport (representing a "rural" setting), he found that the city was appreciably warmer (3–3.5°C or 5.4–6.3°F) and drier (relative humidity up to 20 percent lower) than the airport (Figure 2.5a). The narrow

(a)

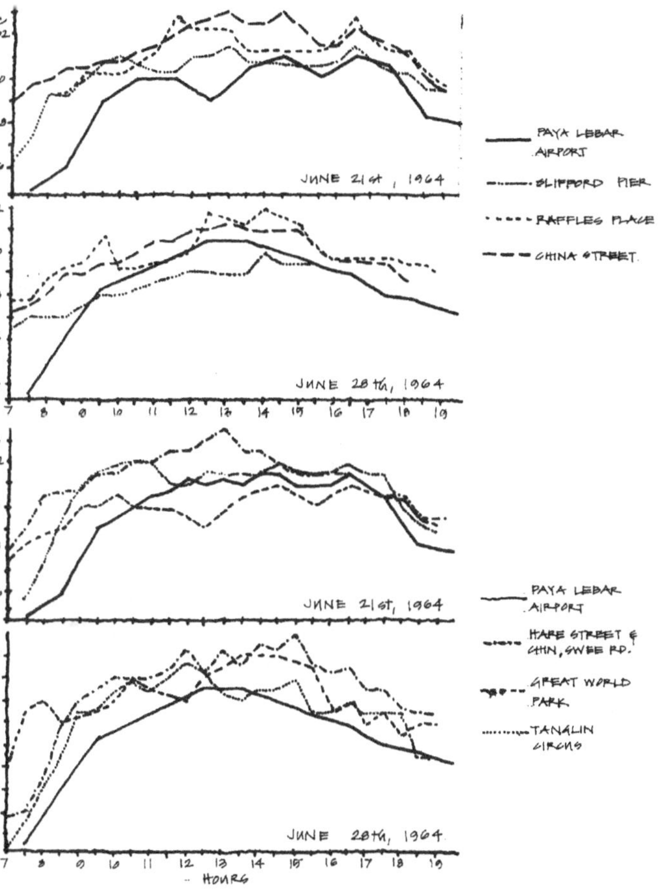

(b)

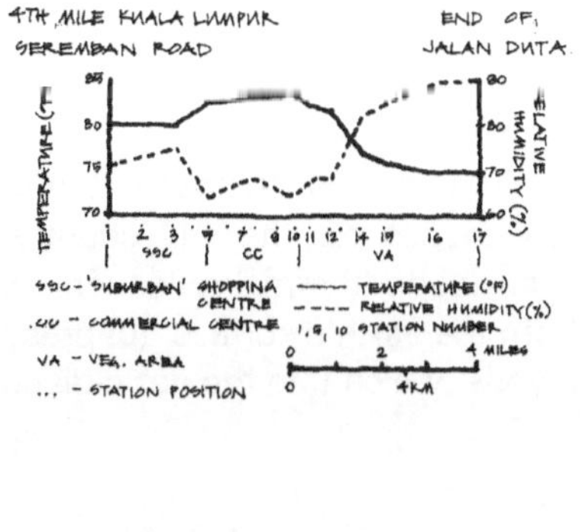

2.5 *UHI manifestations in Singapore and Kuala Lumpur.*
Sources: (a) Nieuwolt (1966); (b) Sani (1973)

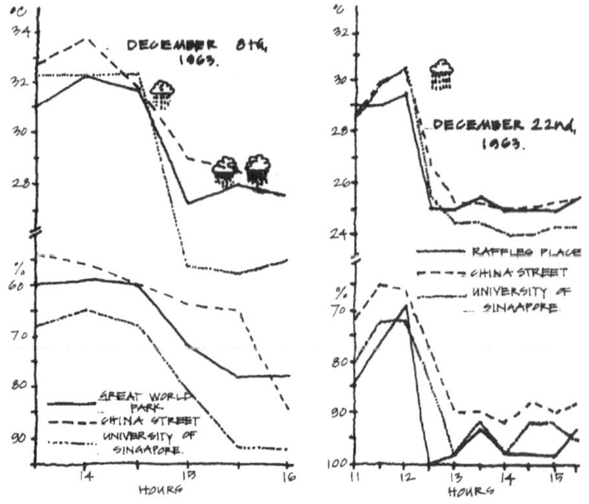

2.6 *Thunderstorms and tropical UHIs.*
Source: Nieuwolt (1966)

streets in the central parts of the city had the worst microclimate. During the day, the sea front was wetter and cooler than the rest of the city, but the effect was found only on a very narrow strip of land around the sea (due to tall buildings blocking the benevolent effects of wind).

Interestingly, the effect of wind was found to *increase* the comfort differences between the city and airport, to the detriment of the former. Nieuwolt attributed the climatic differences to the increased absorption of solar radiation by the city and its lack of evapotranspiration, with a greater emphasis on the latter. (In the rural tropics evaporation accounts for 85 percent of the heat loss [Jen Hu Chang, 1965].)

Nieuwolt also found that with the onset of a thunderstorm (a very frequent occurrence in the tropics), air temperature dropped by more than 7°C (12.6°F) (Figure 2.6). In the tropical context where the daily temperature swing is only about 5°C (9°F), the temperature drop associated with the onset of a thunderstorm is indeed very significant.

2.7 *City–country temperature differences in the tropics.*
Source: Sani (1973)

Another tropical UHI study of interest was conducted by Sani (1973) in Kuala Lumpur, Malaysia. His findings agreed with those of Nieuwolt's. However, he also found that the city–country temperature difference (CCTD) was related to cloud cover (Figure 2.7). On clear days the CCTD was 4.4–5°C (8–9°F), but on cloudy days it was only about 2–2.2°C (3.54°F). He attributed the CCTD to surface characteristics of the city (more dark-colored surfaces) and the lack of vertical air mixing that keeps hot air trapped at body level.

The work of Adebayo (1987, 1990, 1990/91, 1991) in Ibadan, Nigeria, is indicative of UHI studies in tropical Africa. Observing a higher air temperature in the city center than in the rural area, Adebayo cited the following as causes for the city–country temperature difference: (a) reduction in albedo (thermal reflection) which causes more heat to flow to and from the earth's surface, (b) presence of a pollution veil, and (c) urban canyons, which trap the net radiation (Figure 2.8).

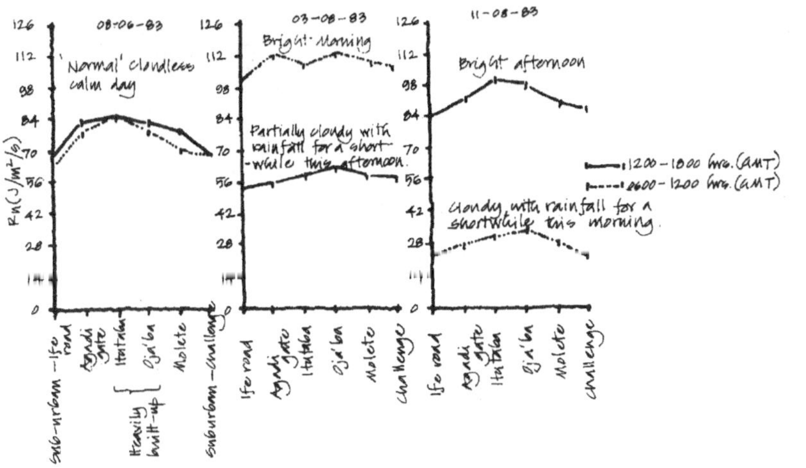

2.8 *Urban radiation balance in the tropics: Ibadan, Nigeria.*
Source: Adebayo (1990)

An equatorial UHI study was recently concluded by Emmanuel (1999a). Figure 2.9 shows the 30-year ambient air temperature trends in the Colombo Metropolitan Region (CMR), the capital city region of Sri Lanka. Since the UHI phenomenon is best seen in the night-time records rather than in the daytime, the graph depicts the daily (diurnal) variation in temperature. A diminishing diurnal temperature variation would indicate a growing UHI problem.

Figure 2.9 indicates the presence of an urban heat island (UHI) in the CMR, with Colombo city having the smallest diurnal variation. This is because of an increase in the daily minimum temperature, which in turn leads to smaller variation in daily temperature. As discussed in the previous section, the daily maximum temperature remains unaffected by urbanization.

But the annual trends in diurnal variations in the CMR are minimal. If one looks at the temperature records alone, there is no indication that the problem is growing, even though the presence of a UHI is

2.9 *Historic air temperature trends in the CMR*

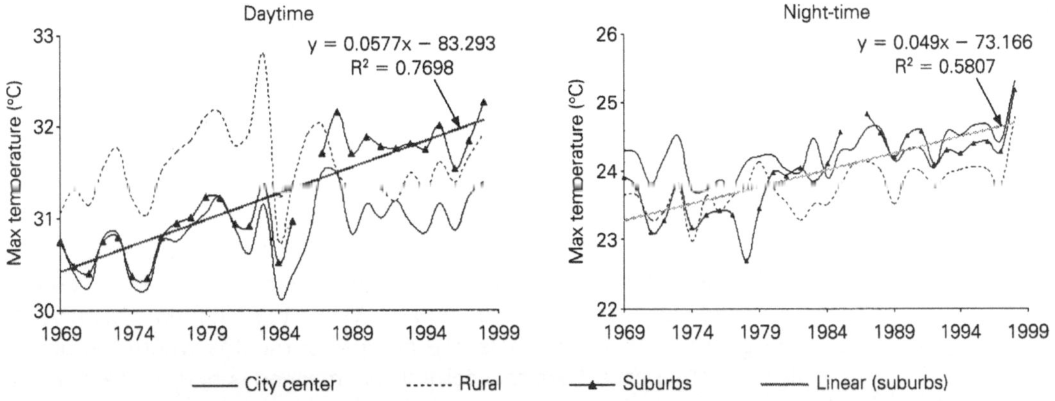

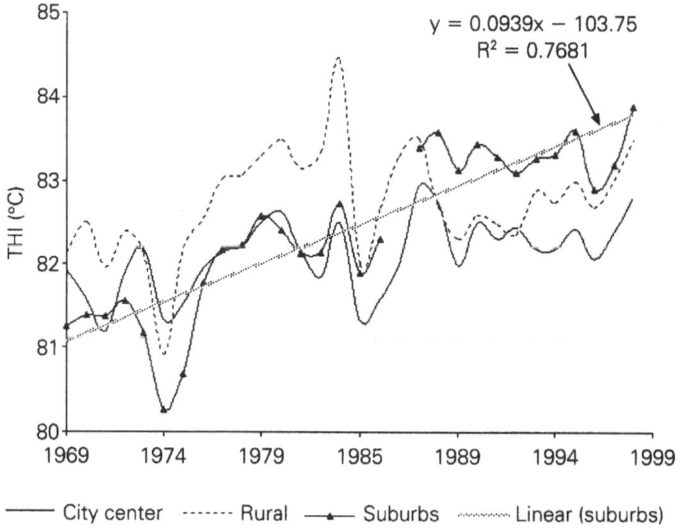

2.10 *Daytime thermal comfort trends in the hottest month*

confirmed. However, this is true only if air temperature records are considered in isolation. The picture is different if thermal comfort trends, instead of air temperature, are compared for the CMR.

Figure 2.10 shows the daytime thermal comfort trends in the region during the last 30 years (average annual trends and the trends during the hottest month – April – respectively). The thermal comfort trends are measured in terms of a combined comfort index called the temperature–humidity index (THI) (Deosthali, 1999). (See Chapter 3 for detailed discussion.)

Two conclusions can be made from Figure 2–10. Much like the air temperature regime, thermal comfort in the city is better than in the rural surroundings during the daytime. This conforms to previous research findings. For example, Plumley (1977) found that a city's thermal comfort condition is better during daytime due to the shading effects of buildings.

The second conclusion is that the recent climate of the CMR is beyond the limits of thermal comfort. Even the lowest THI in the region (28.3°C in Colombo city during August) is above the lower limit of thermal discomfort (said to be 26°C).

Figure 2.10 is in sharp contrast to Figure 2.9 where no annual trend is detected. A clear thermal comfort trend is visible in Figure 2.10. This is particularly true for the suburban station, where a statistically significant simple linear regression can be fitted ($R^2 = 0.6008$).

Figure 2.11 shows the night-time thermal comfort patterns in the CMR. Here, too, the presence of a UHI is clear. The city's THI values are higher than those of the suburbs. While the night-time conditions are generally more favorable, the number of months where the upper limit of 100 percent thermal comfort is breached is high in the city (six

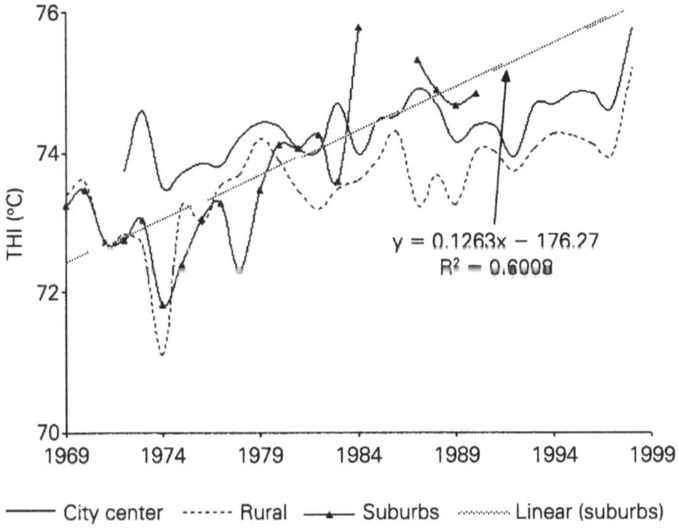

2.11 *Night-time thermal comfort trends in the hottest month*

months in Colombo as opposed to three or four months in the suburbs). Here, too, an annual trend is visible – particularly for the suburban location. However, the night-time annual trend in THI is weaker than that of the daytime.

Figure 2.12 shows a thermal comfort comparison for Colombo city under "typical" and "recent" climate. "Typical" climate data was taken from the period 1920 to 1979. The assumption here is that the "typical" climate encapsulates the "pre-urban" climate conditions in the city. This assumption is valid because rapid urbanization in Colombo occurred only with the introduction of an open economic system in 1977. "Recent" climate is assumed to be from 1994 to 1998. The data from this period indicate the effects of approximately twenty years of rapid urbanization in Colombo.

It appears that Colombo's daytime typical climate was barely tolerable during most of the year (hovering around the upper limit of 100 percent discomfort; that is, THI = 26°C), but the urban changes in the city has made it much more uncomfortable. (The lowest THI value in the recent years is 1.7°C above the lower limit of 100 percent discomfort; see Figure 2.12a.) As for the night-time, the typical climate was within the 100 percent comfort limit almost the whole year, while the urban changes have resulted in approximately 6 months (mid-March to late September) having only a 50 percent comfort level (Figure 2.12b).

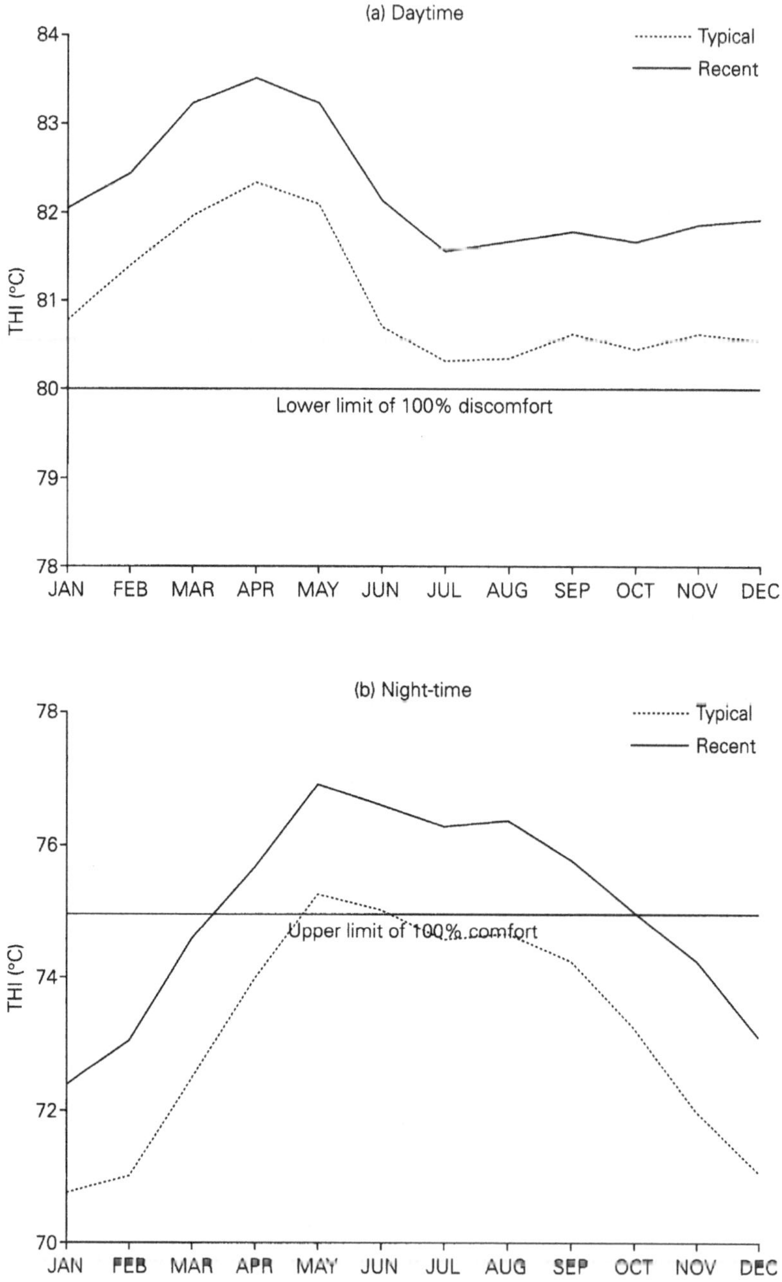

2.12 *"Typical" vs. "recent" climate in Colombo city*

2.4 Urban design variables and microclimate mitigation

There are five urban microclimate features characteristic of the city climate everywhere. First, the UHIs at day and night are very different. A downtown-centered UHI exists at night while the daytime records a mix of cool and heat islands. The magnitude of the temperature differences wanes as background climate becomes hotter. At the same time, hot conditions lead to larger intra-urban thermal comfort differences (see Emmanuel, 1997b). This peculiarity was first noted by Herrington and Vittum (1977), who found a strong correlation between urban thermal comfort and land cover variation, particularly during warm days rather than on cooler ones.

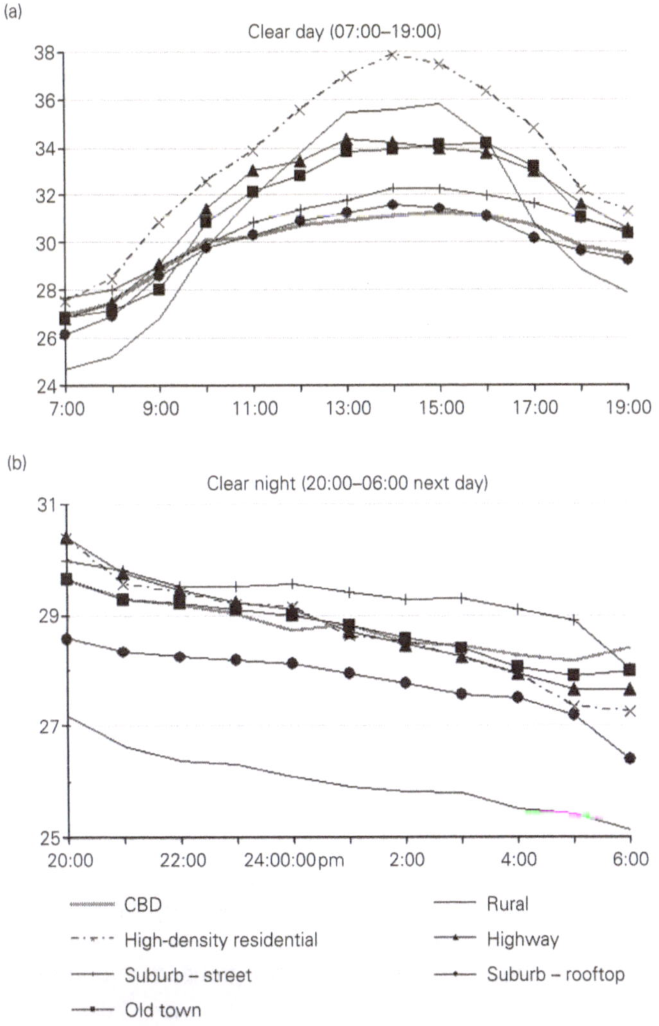

2.13 *Intra-urban thermal variations in the tropics: Colombo, Sri Lanka*

Second, the daytime microclimate variations are strongly correlated to the amount of canyon shading (Figure 2.13). Plumley (1977) recognized this 27 years ago when she noted the importance of shading for daytime urban thermal comfort during warm days. Depending on the amount of shading available, it is possible to create several "cool islands" in the city during daytime. Such a phenomenon in relation to surface temperature patterns was noted by Outcalt (1972b) in a swath of land in northwest Ann Arbor, Michigan. Outcalt found that the tree- and building-shaded old residential quarter was the coolest (the "cold doughnut") and the more open, new subdivisions to be the warmest urban locations. Although it is generally held that the air and surface temperature patterns in urban areas differ substantially (see Eliasson, 1990/91), Nichol (1996a, 1996b) found that these two parameters are similar in the tropics on account of its high solar zenith angles.

The third characteristic relates to the nature of nocturnal UHI. City-wide climate traverse analysis has shown that the downtown-centered UHI at night is highly correlated with urban tree cover characteristics. However, model simulations at the street canyon level reveal that significant night-time cooling could be achieved by increasing the sky-view factor (SVF) and by improving the thermal properties. The importance of tree cover characteristics to *city-wide* night-time UHIs is confirmed by numerous urban climatologists, including Nieuwolt (1966) in Singapore, Sani (1973) in Kuala Lumpur, Malaysia, Gallo *et al.* (1993a, 1993b) for several US cities, and Kawashima (1994) in Tokyo. The relationship between SVF and nocturnal cooling at street level was noted by Barring *et al.* (1985) and Eliasson (2000) in Malmo, Sweden; Swaid (1993b) in Jerusalem; Gustavsson (1995) in Halland county, Sweden, and by Shashua-Bar and Hoffman (2000, 2002a, 2002b) in Tel Aviv.

The fourth feature of UHI relates to the utility of extensive tree cover in the street canyon. Thermal comfort comparisons at a completely vegetation-covered site show that the daytime cooling associated with extensive tree canopy was not very different from cooling by building shade (see, for example, Nichol, 1996a; Souch and Souch, 1993). Furthermore, extensive tree cover inhibited night-time comfort due to the reduced SVF.

Finally, the urban climate modifications at the downtown areas are very different to those at the peripheral areas of cities. A possible reason for this difference is the sharp contrast in street canyon geometry between the downtown and the periphery ("a very dense core and scattered periphery": Oke, 1988b: 111). This suggests that UHI mitigation strategies for these two areas need to be different.

In light of these findings we can now evaluate the most commonly suggested UHI mitigation strategies, particularly from a tropical thermal comfort point of view (Table 2.4).

The UHI mitigation strategies can be grouped into three major themes:

Table 2.4 Optimal urban parameters suggested for mitigating the negative impact of urban heat island (UHI)

Urban parameters	*Optimal values*
Canyon Geometry	Height: width ratio of 0.4-0.6 is suggested by Oke (1988b) for minimal heat trapping in summer and enhanced trapping in winter, without infringing too much upon air quality standards.
Thermal properties	An albedo increase of 0.15 in Sacramento, California, was shown (Sailor, 1995) to have reduced the UHI by 2.7°F.
Moisture availability	The average moisture availability in North American cities is 15 percent of that in rural areas (Monteith, 1973). Its doubling was found to reduce the UHI by 20 percent in a simulation study (Oke *et al.*, 1991).
Anthropo-genic heat	Sailor (1995) suggested better building insulation and compact urban planning for a 50 percent reduction in building and urban heat wastes.
Vegetative cover	Average North American cities are about 30 percent tree covered (Moll and Ebenreck, 1989). Sailor (1995) found that a doubling of this amount could reduce the UHI by about 2°F.

1. Increase vegetation cover.
2. Increase thermal reflectivity (albedo) of urban surfaces, particularly roofs.
3. Manipulate urban geometry.

Increase vegetation cover

The main proponent of this theme is the Heat Island Research Group of the Lawrence Berkeley National Laboratory (LBL), Berkeley, California. Its simulation of summertime UHI suggests that planting trees on at least half the available open spaces in US cities will lead to 15–35 percent reductions in building cooling energy costs nationwide (Akbari *et al.*, 1992). Other studies on climate enhancement/building energy reduction associated with urban vegetation cover have indicated widely varying benefits. These include: up to 27°F (15°C) reduction in surface temperature during hot, summer days in Miami, Florida (Parker, 1983); up to 25 percent reduction in building cooling energy needs in the eastern United States (Heisler, 1986b); 30 percent reduction in daily cooling energy need, and 27–42 percent reduction of peak power demand in Sacramento (Akbari *et al.*, 1997a); 10°F (5.5°C) air temperature reduction within a two-mile radius of Mexico City's urban parks (Jauregui, 1973), and 3–5 percent reduction in regional summertime cooling load for US cities located between 25–45°N latitude with low-to-moderate humidities (Sailor, 1998). On the other hand, Simpson and McPherson (1998) found much smaller cooling load reduction associated with residential trees during the peak cooling load period in Sacramento, California (up to 7.1 percent reduction).

The direct effect of vegetative cover upon urban comfort is not so

Table 2.5 Effect of vegetative cover on heat partitioning.

Green:built	*Rural*	*Suburban*	*Urban*	*Urban center*
	100:0	50:50	15:85	0:100
Energy (Wm^{-2})				
$Q^* + Q_F$	535	554	546	530
Q_{HO}	150	216	240	370
Q_E	305	216	158	0
ΔQ_S	80	122	148	160
Heating rate (k/w)				
Sensible heating	0.5	0.8	0.9	1.3
Evaporative suppression	0.9	0.7	0.5	0
Net change	−0.8	−0.6	−0.5	–

Source: Oke (1989: 343)

Notes:
Q^* = Net radiation
Q_F = Anthropogenic heat
ΔQ_S = Surface heat storage
Q_{HO} = Surface turbulent sensible heat
Evaporative suppression = Thermal equivalent of energy used in evaporation which would otherwise contribute to turbulent warmth.
Net change = Difference from bare city case.

much due to the reduction in air temperature as radiation reduction (Robinette, 1973; Barradas *et al.*, 1999). Trees affect urban microclimates on two levels: human comfort and building energy budget (Parker, 1983; Miller, 1988). It is not only what vegetation does that is important, but also what it prevents – i.e. heating up of urban canyons (Oke, 1989) (Table 2.5). While the direct temperature reduction due to the presence of a large urban park is only about 1–2°C, the prevention of heat build-up (Oke, 1989) and the partitioning of more heat into latent rather than sensible means (see Barradas *et al.*, 1999) will be of greater value to urban dwellers.

However, the ability of urban trees to improve the thermal comfort conditions in the surroundings is a function of the seasons, background climate, size of green area, type of surface over which trees are planted, and the amount of leaf cover. In sub-tropical Mexico City, Barradas *et al.* (1999) found that during the dry season trees dissipate nearly 70 percent of the net radiation via sensible heating and only 25 percent through latent means. In the wet season, however, the numbers are reversed. Thus, water availability is a crucial factor in the increase of latent heat transfer by vegetation. Similarly, cooling effects of urban trees are more pronounced in an area with warmer background climate (Shashua-Bar and Hoffman, 2000). Furthermore, the bigger the urban green pocket, the more pronounced will be the thermal comfort improvement (Gomez *et al.*, 2001). Additionally, trees planted over vegetated areas are able to transpire better than those over asphalt areas (Kjelgren and Montague, 1998). When urban trees completely cover the sky dome, air temperature reductions of up to 3.3°C can be expected (Shashua-Bar and Hoffman, 2002b)

In addition to improving the climate conditions, urban trees also improve the surface water flow by the quantitative and qualitative regulation of runoff (Rutter, 1972; Shuttleworth, 1989), sequester CO_2 (Nowak and McBride, 1993) and cost-effectively reduce other trace gas pollutants (McPherson *et al.*, 1998), enhance human thermal comfort (Heisler, 1974; Stark and Miller, 1977), control smog (Raloff, 1990), add to the psychological well-being of urban dwellers (Ames, 1980; Ulrich, 1984; Willeke, 1989), and enrich urban bio-diversity (Hough, 1984).

Another variant of the urban green enhancement approach is the greening of roof tops (the so-called "green roof" approach). The city of Tokyo is a leading champion of this approach to mitigate the negative effect of the UHI. In the context of a rapid increase in the Tokyo UHI and increasing mortality associated with heat stroke (see Nakai *et al.*, 1999), the Tokyo city authorities have imposed building regulations that mandate at least 20 percent green cover on all newly built structures (see Tanikawa, 2002).

In an experimental study using analogous models, Takakura *et al.*, (2000) found that a concrete roof with ivy cover resulted in a room temperature of 24–25°C while a bare concrete roof resulted in a room temperature of nearly 40°C. However, the night-time cooling experienced by the bare concrete roof was slightly better than that of the ivy-covered roof.

Niachou *et al.* (2001) measured the indoor air temperatures and thermal performance of a roof with and without green cover in a real building in Athens, Greece, during the summer, and then estimated the cooling energy saving due to the green roof. It is reported that green roofs reduced cooling load by up to 48 percent in non-insulated buildings with night ventilation. However, the cooling load reduction due to the green roof in a well-insulated building was negligible (less than 2 percent).

In addition to reducing indoor temperatures and building-level cooling energy use, green roofs also improve the surrounding air temperature on account of a large amount of heat partitioning into latent means (Ishihara and Chou, 1992; Papadakis *et al.*, 2001). However, the life-cycle cost of a green-roof system needs to be taken into account when measuring the total benefits (see Wong *et al.*, 2002). This is particularly so on account of the fact that a well-insulated building achieves similar cooling load reduction to that of a green roof and at a much less operational cost.

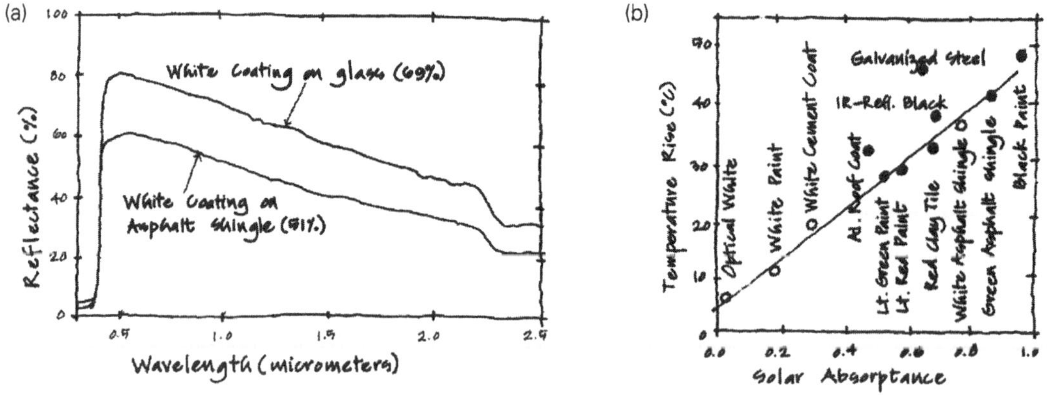

2.14 *Effect of "cool roofs": silver vs. white-washed.*
Source: Akbari *et al.* (1997b)

Increase thermal reflectivity

Proponents of increased urban thermal reflectivity (albedo) encourage light-colored building surfaces, particularly the roofs. Garbesi *et al.* (1989), Akbari *et al.* (1992) and Fishman *et al.* (1994) suggest that increasing urban albedo by up to 15 percent will reduce air temperatures at the neighborhood level by as much as 5°F (2.8°C).

Parker and Barkaszi (1997) white-painted roofs of several actual residential buildings in southern Florida during mid-summer and compared the cooling energy consumption using a before and after protocol. White roofs reduced the average cooling energy consumption by 22 percent (and peak power consumption by 19 percent). Using a similar methodology, Akbari *et al.* (1997b) reported a cooling load reduction of up to 80 percent for a small house and a 35 percent reduction for a large bungalow in Sacramento, California.

Using scale models, Simpson and McPherson (1997) reported slightly better energy consumption performance under a white roof than a silver-colored roof, indicating the importance of emissivity in addition to albedo. This is clearly indicated in Figure 2.14 where silver-colored roof finishes such as galvanized steel and aluminum-coated roofs show greater temperature rise than white-colored roofs (see Berdahl and Bretz, 1997).

Manipulate urban geometry

Of the three main UHI mitigation strategies it is the urban geometry manipulation that has the greatest potential for comfort enhancement at the *neighborhood* scale. The primary design strategy is to explore the shading potential of the urban mass. It may be noted that sun control of individual buildings in the tropics, where the sun is closer to the azimuth, is relatively easier. It is therefore necessary to enhance

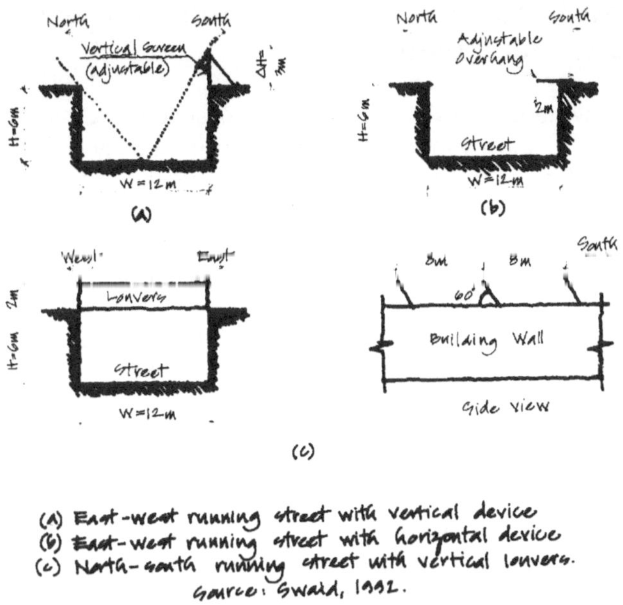

2.15 *Adjustable urban shading devices.*
Source: Swaid (1992)

the comfort potentials of the urban outdoors in the tropics where considerable living occurs. Such an action needs to be taken at the neighborhood scale that utilizes its urban geometry (massing) to reduce the heat-island effect, by shading itself.

Oke (1988b) was one of the first to suggest that improvement to urban geometry at street level needs to (a) reduce UHI in summer, (b) retain heat during winter, and (c) facilitate adequate air change for comfort and pollution dispersal. In order to achieve all three objectives, Oke (1988b) suggested a compromise building height to street width (H:W) ratio of 0.4 to 0.6.

A variant of the geometry manipulation approach was suggested by Swaid (1992) (Figure 2.15). In this proposal an adjustable vertical shading device is attached to the top of a canyon wall. This in turn increases the street canyon depth at daytime (i.e. leads to more shading). The device can be retracted at night, thus the night-time SVF remains unchanged.

An urban approach to shade enhancement was first proposed by Emmanuel (1993). The concept, called "shadow umbrella," utilizes the urban massing to shade the areas between buildings. Chapter 4 details the procedure and presents design strategies derived from this concept.

Sky-view factor as an important tool in urban geometry manipulation

The sky-view factor (SVF) (ψ_s) is a geometrical concept that describes the fraction of the overlying hemisphere occupied by the sky (Oke, 1981: 245). Since the view of the sky is critical for long-wave energy loss (as well as short-wave energy gain), it goes without saying that SVF is of critical utility to urban energetics. Urban climatology has made extensive use of SVF since its introduction to UHI studies by Oke in 1981.

Energetically, SVF is defined as the "ratio of radiation received by a planar surface from the sky to that received from the entire hemispheric radiating environment" (Watson and Johnson, 1987: 193).

Consider a point at the mid-width of the floor of a canyon with symmetric cross-section and infinite length (similar to Figure 2.16). This point will see the building on the left, building on the right, a strip of sky in front (at the far front end of the street), a strip of sky at the back (at the far back end of the street) *and* the strip of sky overhead. Therefore, the SVF is 1 minus the wall views of left wall and right wall. In other words, each thing that the point "sees" has a view factor and each of these could be calculated.

The wall-view factor (ψ_w) represented by each canyon side as viewed from this point can be shown to be:

$$\psi_w = 0.5(\sin^2\theta + \cos\theta - 1)(\cos\theta)^{-1} \qquad (1)$$

Where $\theta = \tan^{-1}(H/0.5W)$.
Therefore the SVF for this point is:

$$\psi_s = (1 - 2\psi_w) \qquad (2)$$

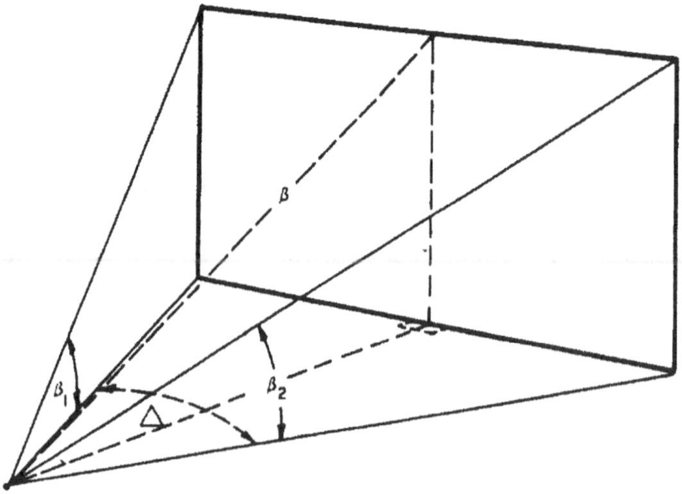

2.16 *SVF calculations: graphical method.*
Source: Watson and Johnson (1987)

For urban canyons in which buildings are irregularly aligned and of finite length, equations (1) and (2) can be refined thus (see Johnson and Watson, 1984):

$$\psi_w = \frac{1}{2\pi}\{(\gamma_2 - \gamma_1) + \cos\beta[\tan^{-1}(\cos\beta\tan\gamma_1)\tan^{-1}(\cos\beta\tan\gamma_2)]\} \quad (3)$$

Where γ_1 and γ_2 are the azimuth (orientation) angles of the wall from the point of interest (where 1 and 2 indicate the closest and the furthest wall ends, respectively), and β is the elevation angle at the mid-point of the wall.

SVF in urban environments may be determined by one of the following three methods (see Watson and Johnson, 1987):

1. Analytical methods: where SVF is calculated once angles from the planar surface to the tops and sides of surrounding are known. The above equations point to this method (nos. 1 and 2 for a symmetrical canyon and eqn. no. 3 for non-symmetrical canyons).
2. Photographic method: a fish-eye lens photograph is used to project the hemispherical radiating environment onto a circular image plane. By overlaying a transparent polar coordinate graph paper on the photograph, the relative fractions of sky and buildings can be counted. Steyn (1980) popularized this approach.
3. Video imagery: a video camera equipped with a fish-eye lens is digitized and analyzed to distinguish the relative areas covered by sky and buildings.

An excellent review of these methods and their derivatives are given in Grimmond *et al.* (2001). However, in order to avoid much confusion and simplify the process, we strongly suggest that the reader uses Watson and Johnson's (1987) graphical method.

Watson and Johnson (1987) give simple-to-read nomograms to calculate wall-view factors. If the elevation angles of both ends of a wall are the same (see Figure 2.16 for angle definitions), then the nomogram shown in Fig. 2.17 should be used. Here the abscissa is the difference in azimuth angles (Δ) and the curves represent the wall elevation angle (β). View factor (ψ_o) on the ordinate is the same as wall-view factor (ψ_w) in this case.

If the elevation angles are different, then:

$$\psi_w = \psi_o - \alpha^*\delta\beta^*\beta_1{}^*\Delta \quad (4)$$

Where ψ_o is obtained from Figure 2.17 as above, and the co-efficient α is read from Figure 2.18 by interpolating between curves representing values of β_1. $\delta\beta$ is the difference between β_1 and β_2.

In both cases (of equal and unequal elevation angles), the sum of ψ_w values subtracted from 1 gives the SVF.

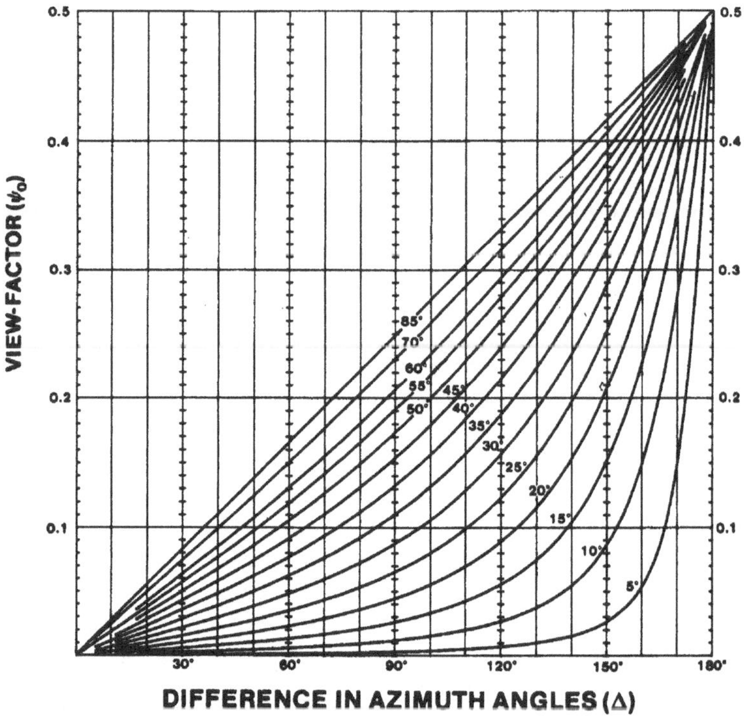

2.17 *SVF nomogram: equal altitudes.*
Source: Watson and Johnson (1987)

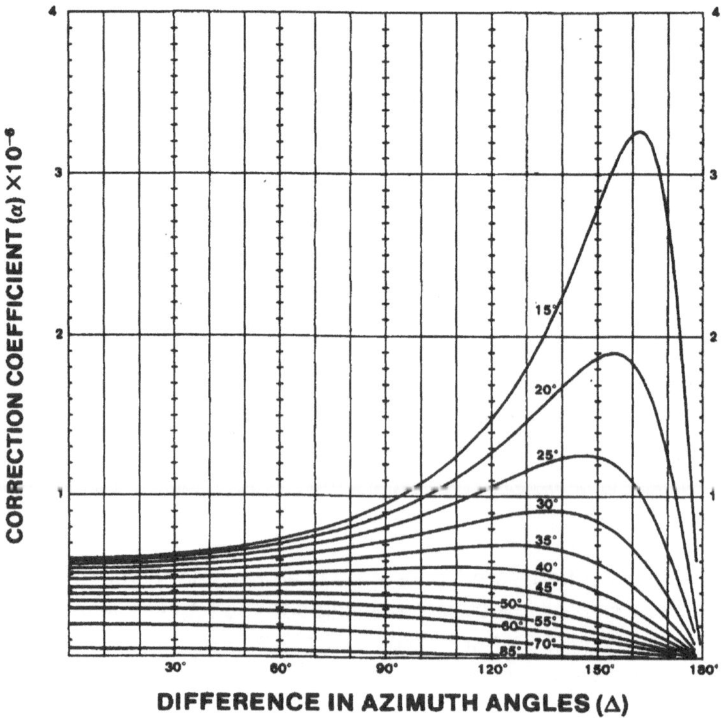

2.18 *SVF nomogram: unequal altitudes.*
Source: Watson and Johnson (1987)

UHI mitigation strategies: a critique

UHI mitigation ought to be viewed as an important design consideration in achieving urban sustainability. The consensus among urban designers concerned with sustainability is that city form needs to facilitate *compact living* (Barton *et al.*, 1995; Moughtin, 1996) for the general improvement of the environment at the macro-scale. (However, there is some evidence that compact living may cause ecological stress at the micro scale – see Whitford *et al.*, 2001.) Golany (1996) sets out the rationale for a compact city form:

> Based on the study of historical practice, we can summarize the advantages of the compact city as follows: respond to and ease the problem posed by stressful climate; consume less energy for cooling and heating; reduce cost of operation of all infrastructure networks; save time and energy of commuting . . . have minimal impact on the delicately sensitive environment. In short, the compact city form responds favorably in its thermal performance to regions of stressful climate (p. 459).

From the point of view of urban sustainability, low canyon geometries fall short of the ideal of compact urban living. The existing canyon geometry at downtown areas of many cities is already within the "optimum" quoted above. Yet a strong UHI persists in these cities. Adopting low canyon geometries will effectively preclude new building activity in many cities, which in turn will exacerbate the urban sprawl.

A general problem with the shade-enhancement approach (urban geometry manipulation as well as shade trees) is that they prevent night-time cooling on account of the restricted sky view (see Emmanuel, 1997c). In this regard, the utility of Swaid's (1992) proposal (see Figure 2.15) appears promising. This adjustable geometry proposal increases street-level shading during the day without compromising the SVF at night, thus ensuring day and night cooling. It is also compatible with the ideal of compact living in that no reduction in building density is warranted.

As for the suggestion regarding urban vegetation enhancement as the mainstay of UHI mitigation, six shortcomings may be pointed out. First, it is necessary to distinguish between climate improvement and energy-need reduction at the *building scale* from those at the *urban scale*. Many authors analyze climate and energy consumption at the building scale (Heisler, 1980a, 1980b, McPherson *et al.*, 1988) and project savings for entire cities (Huang *et al.*, 1987; McPherson *et al.*, 1989; Akbari *et al.*, 1992). While it is true that vegetation cover can lead to significant temperature reductions at the building scale, the urban effects may not be as dramatic. For example, Souch and Souch (1993) reported that there was very little air temperature improvement (less than 2°F) due to vegetation cover at the street canyon scale.

Sailor's (1995) simulation study of the effect of vegetation cover enhancement in the Los Angeles basin region found only about 1.8°F improvement. He simulated vegetation cover increases up to 15 percent in downtown areas, up to 20 percent in multi-family residential neighborhoods and 50 percent cover enhancement in the suburbs. Yet improvements to air temperature larger than 0.5°C (1.0°F) lasted only from about noon until 5 p.m.

Another shortcoming of the vegetation cover enhancement approach is that the potential for such action is the greatest in residential suburbs while cooling is needed the most in downtown areas. Setback regulations in residential zones result in large open spaces, and they are usually filled with extensive tree cover. For example, Emmanuel (1997b) found that single family residential sites in a typical small town in North America had a 40–50 percent tree cover while the downtown had less than 10 percent green cover (with the built-up cover being as high as 87 percent).

Third, urban vegetation cover enhancement approaches usually report the air temperature benefits and assume similar thermal comfort advantages will follow. Unarguably, urban thermal comfort is not well studied, but the little evidence that exists suggests that street-level thermal comfort is a function of view factors of cool surfaces (Herrington and Vittum, 1977). In other words, if hot surfaces abound in a street canyon then thermal comfort will suffer. This means that street-level comfort in summer can be enhanced by shading as many street surfaces as possible. Jendritzky and Nübler (1981) summarize the urban thermal comfort patterns thus:

> The lowest predicted mean vote (PMV) values occur in the city center where building density is high whereas the largest heat load is found in the suburbs or in the rural surroundings . . . The PMV distribution differs markedly from air temperature distribution . . . In midday situations . . . air temperature is an unsuitable means to state thermal comfort. The reason for this is the strong influence of radiation conditions: In the suburbs irradiation is so dominant that it compensates even for the effect of the stronger winds. In the inner city, the possibility to avoid irradiation accounts for the low PMV values (p. 323).

While it is true that street trees can also provide shade in the street canyon, they also decrease the SVF, which in turn will lead to discomfort at night. Additionally, trees exchange convective heat to the ambient air, thus somewhat reducing the cooling potential offered by shading (see Shashua-Bar and Hoffman, 2002a).

Fourth, urban vegetation, particularly that in the street canyon, has a potential to increase the biogenic hydrocarbons (the so-called "ozone forming potential" of trees). While climatic benefits may be experienced, some air quality deterioration might be an inadvertent side-effect, especially in warm, humid cities. Care should be taken to plant low emitters of biogenic hydrocarbons (see Taha *et al.*, 1997).

Fifth, urban vegetation enhancement strategies must take into consideration the excessive amount of tree cover needed to achieve the desired level of comfort enhancement as well as the cost of maintaining the trees (see McPherson, 2001). In the polluted context of urban streets, maintenance cost is a problem that cannot be simply ignored.

Sixth, the cooling potential of vegetation is a function of moisture availability (see Onmura, *et al.*, 2001). At low moisture levels stomatal closure reduces the cooling potential of trees and, if combined with high air temperatures, may actually lead to temperature rise. Furthermore, the cooling potential of vegetation is the highest on relatively cool days (the ratio of net radiation under a tree-shaded area to that in the unshaded area increased from 1:10 to 1:2.4 when the air temperature rose from 36 to 39°C; see Papadakis *et al.*, 2001).

Similarly, three difficulties with albedo enhancement as a mainstay of UHI mitigation efforts may be pointed out. First, large amounts of urban cooling will be possible only if extraordinary increases in albedo are instituted. In Sacramento, California, for example, Taha *et al.* (1988) showed that a 62 percent reduction in urban energy consumption and a 44 percent drop in cooling hours were possible for an urban albedo increase from 0.25 to 0.4. However, in order to achieve an average albedo of 0.40 for urban areas, building surface albedo should be increased to 0.60 (normal value is 0.3), streets should have an albedo of 0.50 (conventional macadamized road has an albedo of 0.14), and other urban surfaces should be increased to 0.25 (about double the values of conventional roofs and sidewalks). In practice, an average urban albedo of 0.4 results only when walls, streets and other man-made urban surfaces are all painted white! As Todhunter (1990a) pointed out, the albedo enhancement approach would be effective only if applied over a larger area (say, an entire city).

Such a profusion of light-colored surfaces will lead to increased light reflections, which is problematic at the urban scale. Akbari *et al.* (1992) cite the traditional Middle East practice of painting buildings in light colors as justification for advocating urban albedo enhancement. However, urban fabric in the Middle East is characterized by deep and narrow street canyons where a significant portion of the light-colored surfaces is constantly in shade. In such situations, light-colored surfaces are useful in reflecting the much-needed sunlight deep into the dark street canyons, without causing glare.

Another problem with implementing an albedo enhancement approach is the lack of empirical data at the urban scale. The LBL research on the matter is theoretical in nature (Garbesi *et al.*, 1989; Akbari *et al.*, 1992; Fishman *et al.*, 1994; Sailor, 1995). This is understandable due to difficulties in calculating *urban* albedo values. Some of the known difficulties include the discrete nature of urban surfaces, constantly changing shade patterns caused by street geometry, presence of vegetation and seasonal variations (Brest, 1987). Remote sensing techniques can overcome some of these difficulties, but street-level albedo is still problematic since remote sensors have a poor view of vertical surfaces as opposed to the horizontal surfaces.

Third, strategies to enhance urban albedo ought to consider the practical issues in maintaining light color in an urban context. Due to excessive pollution and soot deposition, maintaining white color exterior surfaces in urban areas is a monumental task. For example, Pio *et al.* (1998) found that black soot deposition from automobiles reduced reflectance of even sheltered outdoor surfaces by 30 percent in 5–8 years in Oporto, Portugal. Simple methodologies to maintain light color surfaces in urban context do not exist at present (Bretz and Akbari, 1997). Paint degradation due to excessive UV radiation poses special problems to light-colored surfaces in tropical cities.

Directions for further research on UHI phenomena

The purpose of studying urban microclimate alterations is to gain the ability to control their negative impacts (see Tuller, 1975). In terms of urban design, this involves deriving design strategies for the mitigation of the negative effects of UHIs. At present, research on UHI mitigation is very limited. Even the scant research on the matter has four major shortcomings which need to be addressed before implementation is possible.

First, there is a need to specify the impact on humans, particularly the thermal comfort aspect, of urban climate altering factors. Urban climate studies typically report changes on one or two climate parameters only (usually surface or air temperature). Such information is insufficient to derive human comfort effects. Thermal comfort is a complex reaction to at least four environmental and two "subjective" variables: air temperature, radiant temperature, air velocity, humidity, clothing levels, and activity types (see Fanger, 1970). Without accompanying information on the human cost of urban microclimate alterations, urban temperature information alone is not helpful to architects and urban designers to make informed design choices.

Second, current literature rarely evaluates design strategies suggested for the mitigation of the urban heat-island effect. Although many urban climate studies do end with design proposals (e.g. Sundborg, 1950; Nieuwolt, 1966; Barring *et al.*, 1985; Eliasson, 1990/91), design strategies are of secondary importance to these studies. Their primary interest is in explaining the energetic/causative aspects of urban climate change. On the other hand, studies which focus on mitigation tend to be theoretical in approach and lack empirical validation (e.g. Golany, 1996; Pressman, 1996).

The third shortcoming is the incompatibility of current design suggestions, such as those by Oke (1988b), with the need for compact urban living. For example, it is widely suggested that urban climate could be improved if building densities could be lowered (to an urban canyon H:W ratio of 0.4–0.6) and vegetation presence increased. But such an approach reinforces the urban sprawl to accommodate the ever-increasing urban population. This is especially the case in tropical cities where urbanization is intense. Low-density development could only bring about climate alterations to an even larger area.

Finally, there is a need for empirical data in UHI research. Experiments comparing urban climate mitigation regimes tend to employ the simulation approach. Controlled experimentation on urban heat islands remains very difficult.

In summary, we can state that the study of urban climate modifications is well advanced. Even by the late 1970s Oke (1979) found well over 3,000 studies on urban climate alterations. During a ten-year period ending in 1991, Jauregui (1993) found 169 studies on tropical urban climate alone. What is lacking, however, are design efforts by urban designers to transform the large body of knowledge into design strategies (Swaid *et al.*, 1993; de Schiller and Evans, 1996). Even the few that exist need to be evaluated on the basis of their thermal comfort effects. The critical need of the urban design community is the quantification of thermal comfort effects of urban parameters that are controlled by urban design.

References

Acharya, P.K. (1979). *Indian Architecture According to Mänasära-Šilpasästra*, Patna, India: Indian India.

Adebayo, Y.R. (1987). "Land-use approach to the spatial analysis of the urban 'heat island' in Ibadan, Nigeria," *Weather*, 42(9): 273–280.

Adebayo, Y.R. (1990). "Aspects of the variation in some characteristics of radiation budget within the urban canopy of Ibadan," *Atmospheric Environment B*, 24(1): 9–17.

Adebayo, Y.R. (1990/91). "Consideration for climate sensitive design in Tropical Africa," *Energy and Buildings*, 15/16: 15–21.

Adebayo, Y.R. (1991). "'Heat Island' in a humid tropical city and its relationship with potential evaporation," *Theoretical and Applied Climatology*, 43(3): 137–147.

Akbari, H., S. Davis, S. Dorsano, J. Huang, and S. Winnett (eds), (1992). *Cooling Our Communities: A Guidebook to Tree Planting and Light-Colored Surfacing*, Washington, D.C.: US Environmental Protection Agency.

Akbari, H., D.M. Kurn, S.E. Bretz, and J.W. Hanford (1997a). "Peak power and cooling energy savings of shade trees," *Energy and Buildings*, 25(2): 139–148.

Akbari, H., S. Bretz, D.M. Kurn, and J. Hanford (1997b). "Peak power and cooling energy savings of high-albedo roofs," *Energy and Buildings*, 25(2): 117–126.

Ames, R.G. (1980). "The sociology of urban tree planting," *Journal of Arboriculture*, 6(5): 120–123.

Arnfield, A.J. (1990). "Canyon geometry, the urban fabric and nocturnal cooling: a simulation approach," *Physical Geography*, 11(3): 220–239.

Baik, J.-J. and H.-Y. Chun (1997). "A dynamic model for urban heat islands," *Boundary-Layer Meteorology*, 83: 463–477.

Balling, R.C. and S.B. Idso (1989). "Historical temperature trends in the United States and the effect of urban population growth," *Journal of Geophysical Research*, 94(D3): 3359–3363.

Barradas, V.L., A. Tejeda-Martinez, and E. Jauregui (1999). "Energy balance measurements in a suburban vegetated area in Mexico City," *Atmospheric Environment*, 33(24–25): 4109–4113.

Barring, L., J.O. Mattsson, and S. Lindqvist (1985). "Canyon geometry, street temperatures and urban heat island in Malmo, Sweden," *Journal of Climatology*, 5: 433–444.

Barr-Kumarakulasinghe, S.A. (1997). "Monthly temperature trends in South India and Sri Lanka," http://www.xdc.arm.gov/slclimate/temp_trend.html

Barton, H., G. Davis, and R. Gusie (1995). *Sustainable Settlements: A Guide for Planners, Designers and Developers*, Bristol, UK: University of West England.

Berdahl, P. and S.E. Bretz (1997). "Preliminary survey of the solar reflectance of cool roofing materials," *Energy and Buildings*, 25(2): 149–158.

Bornstein, R.D. (1975). "The two-dimensional URBMET urban boundary layer model," *Journal of Applied Meteorology*, 14: 1459–1477.

Bosselmann, P., E. Arens, K. Dunker, and R. Wright (1995). "Urban form and climate: case study, Toronto," *Journal of the American Planning Association*, 61(2): 226–239.

Brest, C.L. (1987). "Seasonal albedo of an urban/rural landscape from satellite observations," *Journal of Climate and Applied Meteorology*, 26(9): 1169–1187.

Bretz, S.E. and H. Akbari (1997). "Long-term performance of high-albedo roof coatings," *Energy and Buildings*, 25(2): 159–167.

Caldwell, C. (1798). *A Semi-Annual Oration, on the Origin of Pestilential Diseases*, Philadelphia, Pa.: Thomas and Samuel F. Bradford.

Caldwell, C. (1833). "Thoughts on the means of protecting health in hot climates," *Transylvania Journal of Medicine & Associated Sciences*, 6: 357–377.

Caldwell, C., (1836). "Thoughts on hygiene as applicable to hot climates, more especially to the Mississippi region," *Transylvania Journal of Medicine & Associated Sciences*, 9: 601–641.

Carlson, T.N., J.K. Dodd, S.G. Benjamin, and J.N. Cooper (1981).

"Satellite estimation of surface energy balance, moisture availability and thermal inertia," *Journal of Applied Meteorology*, 20: 67–87.

Chandler, T.J. (1965). *The Climate of London*, London: Hutchinson.

Chandler, T.J. (1976). *Urban Climatology and its relevance to Urban Design*, WMO Technical Note No. 149, Geneva: World Meteorological Organization.

Currie, W. (1792). *An Historical Account of Climates and Diseases of America*, Philadelphia, Pa.: T. Dobson.

de Dear, R.J. and J.C. Potter (eds) (1999). *Proceedings of the 15th International Congress of Biometeorology and International Conference on Urban Climatology*, Sydney, Australia: Macquarie University.

de Schiller, S. and J.M. Evans (1996). "Training architects and planners to design with urban microclimates," *Atmospheric Environment*, 30(3): 449–454.

Deosthali, V. (1999). "Assessment of impact of urbanization on climate: an application of bio-climatic index," *Atmospheric Environment*, 33(24–25): 4125–4133.

Eliasson, I. (1990/91). "Urban geometry, surface temperature and air temperature," *Energy and Buildings*, 15/16: 141–145.

Eliasson, I. (2000). "The use of climate knowledge in urban planning," *Landscape and Urban Planning*, 48: 31–44.

El-Sioufi, M.M., (1987). "Urban patterns for improved thermal performance," Darch. dissertation, University of Michigan.

Emmanuel, R. (1993) "A hypothetical 'shadow umbrella' for thermal comfort enhancement in the equatorial urban outdoors," *Architectural Science Review*, 36(4): 173–184.

Emmanuel, R. (1997a). "Urban vegetational change as an indicator of demographic trends in cities: the case of Detroit," *Environment and Planning B*, 24: 415–426.

Emmanuel, R. (1997b). "Summertime urban heat island mitigation: propositions based on an investigation of intra-urban air temperature variations," *Architecture Science Review*, 40(4): 155–164.

Emmanuel, R. (1997c). "Summertime heat island effects of urban design parameters," Ph.D. dissertation, University of Michigan, Ann Arbor, USA.

Emmanuel, R. (1999a). "Trends in urban climate changes in Sri Lanka: the case of the Colombo Metropolitan Region," Research sponsored by the Senate Research Committee (SRC) of the University of Moratuwa, Grant No. 99/01/11 (available in the University of Moratuwa Library).

Emmanuel, R. (1999b). "Urban heat island and cooling load: the case of an equatorial city," *Architecture, Energy & Environment*, 16:1–16:8.

Emmanuel, R. (2003). "Assessment of impact of land cover changes on urban bioclimate: the case of Colombo, Sri Lanka," *Architectural Science Review*, 46(2): 151–158.

Fanger, P.O. (1970). *Thermal Comfort: Analysis and Applications in Environmental Engineering*, New York: McGraw-Hill.

Fishman, B., H. Taha, and H. Akbari (1994). *Meso-Scale Cooling Effects of High Albedo Surfaces: Analysis of Meteorological Data from White Sands National Monument and White Sands Missile Range*, LBL Report No. 35056, Berkeley, Calif.: Lawrence Berkeley Laboratory.

Gallo, K.P., A.L. McNab, T.R. Karl, J.F. Brown, J.J. Hood, and J.D. Tarpley (1993a). "The use of NOAA/AVHRR data for assessment of the urban heat island effect," *Journal of Applied Meteorology*, 32: 899–908.

Gallo, K.P., A.L. McNab, T.R. Karl, J.F. Brown, J.J. Hood, and J.D. Tarpley (1993b). "The use of a vegetation index for the assessment of the urban heat island effect," *International Journal of Remote Sensing*, 14(11): 2223–2230.

Garbesi, K., H. Akbari, and P. Martien (eds) (1989). *Controlling Summer Heat Islands*, Proceedings of workshop on saving energy and reducing atmospheric pollution by controlling summer heat islands, February 23–24, 1989, Berkeley, Calif. LBL Report No. 27872, Berkeley, Calif.: Lawrence Berkeley Laboratory.

Geiger, R. (1957). *The Climate Near Ground*, translated by M.N. Stewart *et al.*, Cambridge, Mass.: Harvard University Press.

Golany, G.S. (1996). "Urban design morphology and thermal performance," *Atmospheric Environment*, 30(3): 455–465.

Gomez, F., N. Tamarit, and J. Jaboloyes (2001). "Green zones, bioclimatics studies and human comfort in the future development of urban planning," *Landscape and Urban Planning*, 55: 151–161.

Goward, S.N. (1981). "Thermal behavior of urban landscapes and the urban heat island," *Physical Geography*, 2(1): 19–33.

Grimmond, C.S.B. and T.R. Oke (1995). "Comparison of heat fluxes from summertime observations in the suburbs of four North American cities," *Journal of Applied Meteorology*, 34(4): 873–889.

Grimmond, C.S.B., S.K. Potter, H.N. Zutter, and C. Souch (2001). "Rapid methods to estimate sky-view factors applied to urban areas," *International Journal of Climatology*, 21: 903–913.

Gustavsson, T. (1995). "A study of air and road surface temperature variations during clear windy nights," *International Journal of Climatology*, 15: 919–932.

Heisler, G.B. (1974). "Trees and human comfort in urban areas," *Journal of Forestry*, 72: 466–469.

Heisler, G.B. (1986a). "Energy savings with trees," *Journal of Arboriculture*, 12(5): 113–125.

Heisler, G.B. (1986b). "Effects of individual trees on the solar radiation climate of small buildings," *Urban Ecology*, 9: 337–359.

Henry, J.A. and S.E. Dicks (1987). "Association of urban temperatures with land use and surface materials," *Landscape and Urban Planning*, 14(1): 21–29.

Herrington L.P. and J.S. Vittum (1977). "Human thermal comfort in urban outdoor spaces," *Proceedings of a Conference on Metropolitan Physical Environment*, Syracuse, N.Y., 25–29 August, 1975. USDA Forest Service Technical Report NE-25, Upper Darby, Pa.: Northeastern Forest Experiment Station, pp. 130–138.

Hough, M. (1984). *City Form and Natural Processes: Towards a New Urban Vernacular*, London: Croom Helm.

Howard, L. (1833). *Climate of London Deduced from Meteorological Observations*, (3rd edn), London: Harvey and Darton.

Huang, Y.J., H. Akbari, H. Taha, and A.H. Rosenfeld (1987). "The potential of vegetation in reducing summer cooling loads in residential buildings," *Journal of Climate and Applied Meteorology*, 26: 1103–1116.

Ishihara, O. and S. Chou (1992). "Experimental study on environmentally lightening effect of soil and lawn vegetation on roof," *Proceedings of the Annual Meeting of the Japanese Society of Solar Energy*, pp. 255–258.

Jauregui, E. (1973). "Untersuchungen um Stadtklima von Mexico-Stadt," Doctoral thesis, Rheinischen Friedrich-wilhelms-Universitat.

Jauregui, E. (1986a). "Tropical urban climates: a review," in *Urban Climatology and its Applications with Special Regards to Tropical Areas*, Proceedings of Technical Conference, WMO 652, Geneva, Switzerland: WMO.

Jauregui, E. (1986b). "Urban climatology in the tropics," *Report of First Session of CCI Working Group on Climate and Urban Areas Including Buildings and Other Aspects*, WCAP-8, WMO/TD No. 287, Geneva: World Meteorological Organization, pp. 13–19.

Jauregui, E. *et al.* (1992). "Aspects of heat-island development in Guadalajara, Mexico," *Atmospheric Environment*, 26B(3): 391–396.

Jauregui, E. (1993). *Bibliography of Urban Climate in Tropical/Sub-Tropical Areas, 1981–1991*, WMO/TD No. 552, WCASP No. 25, Geneva: WMO.

Jauregui, E. (1996). *Bibliography of Urban Climatology for the Period 1992–1995, Including a Special Section on Urban Climate in Tropical/Sub-Tropical Areas*, WMO/TD No. 759, WCASP No. 36, Geneva: WMO.

Jauregui, E. (1997). "Heat island development in Mexico City," *Atmospheric Environment*, 31(22): 3821–3831.

Jauregui, E. and E. Romales (1996). "Urban effects on convective precipitation in Mexico City," *Atmospheric Environment*, 30(20): 3383–3389.

Jauregui, E. and A. Tejeda (1997). "Urban–rural humidity contrast in Mexico City," *International Journal of Climatology*, 17(2): 187–196.

Jauregui, E., J. Cervantes, and A. Tejeda (1997). "Bioclimatic conditions in Mexico City: an assessment," *International Journal of Biometeorology*, 40: 166–177.

Jen Hu Chang (1965). "On the study of evapotranspiration and water balance," *Erdkunde*, p. 144.

Jendritzky, G. and W. Nübler (1981). "A model analyzing the urban thermal environment in physiologically significant terms," *Archives for Meteorology, Geophysics and Bioclimatology, Series B*, 29: 313–326.

Johnson, G.T. and I.D. Watson (1984). "The determination of view factors in urban canyons," *Journal of Climate and Applied Meteorology*, 2: 329–335.

Kawashima, S. (1994). "Relation between vegetation, surface temperature and surface composition in the Tokyo region during winter," *Remote Sensing of Environment*, 50: 52–60.

Kidder, S.Q. and O.M. Essenwanger (1995). "The effect of clouds and wind on the difference in nocturnal cooling rates between urban and rural areas," *Journal of Applied Meteorology*, 34(11): 2440–2448.

Kjelgren, R. and T. Montague (1998). "Urban tree transpiration over turf and asphalt surfaces," *Atmospheric Environment*, 32(1): 35–41.

Landsberg, H.E. (1956). "The climate of towns," in W.D. Thomas (ed.), *Man's Role in Changing the Face of the Earth*, Chicago, Ill.: University of Chicago Press, pp. 584–606.

Landsberg, H.E. (1981). *The Urban Climate*, New York: Academic Press.

Lewis, J.E. and T.N. Carlson (1989). "Spatial variations in regional surface energy exchange patterns for Montreal, Quebec," *The Canadian Geographer*, 33(3): 194–203.

McPherson, E.G. (2001). "Sacramento's parking lot shading ordinance: environmental and economic costs of compliance," *Landscape and Urban Planning*, 57: 105–123.

McPherson, E.G. *et al.* (1988). "Impacts of vegetation on residential heating and cooling," *Energy and Buildings*, 12: 41–51.

McPherson, E.G., J.R. Simpson, and M. Livingston (1989). "Effects of three landscape treatments on residential energy and water use in Tuscon, AZ," *Energy and Buildings*, 13: 127–138.

McPherson, E.G., K.I. Scott, and J.R. Simpson (1998). "Estimating cost effectiveness of residential yard trees for improving air quality in Sacramento, California, using existing models," *Atmospheric Environment*, 32(1): 75–84.

Manley, G. (1958). "On the frequency of snowfall in metropolitan England," *Quarterly Journal of the Royal Meteorological Society*, 84: 70–72.

Meyer, W.B. (1991). "Urban heat island and urban health: early American perspectives," *Professional Geographer*, 43(1): 38–48.

Miller, R.W. (1988). *Urban Forestry: Planning and Managing Urban Green Spaces*, Englewood Cliffs, N.J.: Prentice-Hall.

Mills, G.M. (1993). "Simulation of the energy budget of an urban canyon – Part I: model structure and sensitivity test," *Atmospheric Environment*, 27B(2): 157–170.

Mills, G. (1997). "An urban canopy-layer climate model," *Theoretical and Applied Climatology*, 57: 229–244.

Moll, G. and S. Ebenreck (eds) (1989). *Shading Our Cities*, Washington, D.C.: Island Press.

Monteith, J.L. (1973). *Principles of Environmental Physics*, London: Edward Arnold.

Morgan, M.H. (trans.) (1960). *Vitruvius: The Ten Books on Architecture*, New York: Dover.

Moughtin, C. (1996). *Urban Design: Green Dimensions*, Oxford: Butterworth Architecture.

Myrup, L.O. (1969). "A numerical model of the urban heat island," *Journal of Applied Meteorology*, 32: 896–907.

Nakai, S., T. Itoh, and T. Morimoto (1999). "Death from heat-stroke in Japan: 1968–1994," *International Journal of Biometeorology*, 43: 124–127.

Niachou, A., K. Papakonstantinou, M. Santamouris, A. Tsangrassoulis, and G. Mihalakakou (2001). "Analysis of the green roof thermal properties and investigation of its energy performance," *Energy and Buildings*, 33: 719–729.

Nichol, J.E. (1996a). "High-resolution surface temperature patterns related to urban morphology in a tropical city: a satellite-based study," *Journal of Applied Meteorology*, 35(1): 135–146.

Nichol, J.E. (1996b). "Analysis of the urban thermal environment with LANDSAT data," *Environment and Planning B*, 23: 733–747.

Nieuwolt, S. (1966). "The urban microclimate of Singapore," *Journal of Tropical Geography*, 22: 30–37.

Nowak, D.J. and J.R. McBride (1993). "Testing microdensitometric ability to determine Monterey Pine urban tree stress," *Photogrammetric Engineering and Remote Sensing*, 59(1): 89–91.

Oke, T.R. (1974). *A Review of Urban Climatology 1968–1973*, WMO Technical Note 134, Geneva: WMO.

Oke, T.R. (1976). "The distinction between canopy and boundary-layer heat islands," *Atmosphere*, 14(4): 268–277.

Oke, T.R. (1979). *Review of Urban Climatology, 1974–1978*, WMO Technical Note 169, Geneva: WMO.

Oke, T.R. (1981). "Canyon geometry and the nocturnal urban heat island: comparison of scale model and field observations," *Journal of Climatology*, 1: 237–254.

Oke, T.R. (1982). "The energetic basis of urban heat island," *Quarterly Journal of the Royal Meteorological Society*, 108 (455): 1–24.

Oke, T.R. (1987). *Boundary Layer Climates*, London: Methuen.

Oke, T.R. (1988a) "The urban energy balance," *Progress in Physical Geography*, 12(4): 471–508.

Oke, T.R. (1988b) "Street design and urban canopy layer climate," *Energy and Buildings*, 11: 103–113.

Oke, T.R. (1989). "The micrometeorology of the urban forest," *Philosophical Transactions of Royal Society of London B*, 324: 335–350.

Oke, T.R. (1991). "Climate of cities," in F. Baer, N.L. Canfield, and J.M. Mitchell (eds), *Climate in Human Perspective: A Tribute to Helmut E. Landsberg*, Dordrecht, Netherlands: Kluwer Academic Publishers, pp. 61–75.

Oke, T.R., G.T. Johnson, D.G. Steyn, and I.D. Watson (1991). "Simulation of surface urban heat islands under 'ideal' conditions at night. Part 2: diagnosis of causation," *Boundary-layer Meteorology*, 56: 339–358.

Oke, T.R., R.A. Spronken-Smith, E. Jauregui, and C.S.B. Grimmond (1999). "The energy balance of central Mexico City during the dry season," *Atmospheric Environment*, 33: 3919–3930.

Onmura, S., M. Matsumoto, and S. Hokoi (2001). "Study on evaporating cooling effect of roof lawn garden," *Energy and Buildings*, 33: 653–666.

Outcalt, S.I. (1972a). "The development and application of a simple digital surface climate simulator," *Journal of Applied Meteorology*, 11: 629–636.

Outcalt, S.I. (1972b). "A reconnaissance experiment in mapping and modeling the effect of land use on urban thermal regimes," *Journal of Applied Meteorology*, 11: 1369–1373.

Papadakis, G., P. Tsamis, and S. Kyritsis (2001). "An experimental investigation of the effect of shading with plants for solar control of buildings," *Energy and Buildings*, 33: 831–836.

Parker, D.S. and S.F. Barkaszi, Jr. (1997). "Roof solar reflectance and cooling energy use: field research results from Florida," *Energy and Buildings*, 25(2): 105–115.

Parker, J. (1983). "Do energy conserving landscape work?," *Landscape Architecture*, 72(4): 89–90.

Pease, R.W., J.E. Lewis, and S.I. Outcalt (1976). "Urban terrain climatology and remote sensing," *Annals of the Association of American Geographers*, 66(4): 557–569.

Peterson, J.T. (1969). *The Climate of Cities: A Survey of Recent Literature*, Raleigh, N.C.: National Air Pollution Control Administration.

Pio, C.A., M.M. Ramos, and A.C. Duarte (1998). "Atmospheric aerosol and soiling of external surfaces in an urban environment," *Atmospheric Environment*, 32(11): 1979–1989.

Plumley, H.J. (1977). "Design of outdoor spaces for thermal comfort," *Proceedings of a Conference on Metropolitan Physical Environment*, Syracuse, N.Y., 25–29 August 1975, USDA Forest Service Technical Report NE-25, Upper Darby, Pa.: Northeastern Forest Experiment Station, pp. 152–162.

Pressman, N.E.P. (1996). "Sustainable winter cities: future directions for planning, policy and design," *Atmospheric Environment*, 30(3): 521–529.

Raloff, J. (1990). "Urban smog control: a new role for trees?," *Science News*, 138: 5 (July 7).

Robinette, G.O. (1973). *Energy and Environment*, Dubuque, Ia.: Kendall/Hunt Publishers.

Rutter, A.J. (1972). *Transpiration*, London: Oxford University Press.

Sailor, D.J. (1995). "Simulated urban climate response to modifications in surface albedo and vegetative cover," *Journal of Applied Meteorology*, 34: 1694–1704.

Sailor, D.J. (1998). "Simulations of annual degree day impacts of urban vegetative augmentation," *Atmospheric Environment*, 32(1): 43–52.

Saitoh, T.S. and H. Hoshi (1993). "Urban warming in Tokyo and counter-plan to improve future environment," *IEEE Transactions* 9338-6: 2.887-2.892.

Saitoh, T.S., T. Shimada, and H. Hoshi (1996). "Modelling and Simulation of the Tokyo urban heat island," *Atmospheric Environment*, 30(20): 3431–3442.

Sani, S. (1973). "Observations on the effect of a city form and functions on temperature patterns," *Journal of Tropical Geography*, 36: 60–65.

Schmidt, W. (1929). Quoted by R. Geiger (1957), *op. cit.*

Shashua-Bar, L. and M.E. Hoffman (2000). "Vegetation as a climatic component in the design of an urban street: an empirical model for predicting the cooling effect of urban green areas with trees," *Energy and Buildings*, 31: 221–235.

Shashua-Bar, L. and M.E. Hoffman (2002a). "Quantitative evaluation of tree effects on diurnal air temperature cooling in urban streets," in ACAD and GRECO (eds), *Design With Environment*, Proceedings of the 19th Passive and Low Energy Architecture (PLEA) Conference, Toulouse, France, July 2002, pp. 235–240.

Shashua-Bar, L. and M.E. Hoffman (2002b). "The green CTTC model for predicting the air temperature in small urban wooded sites," *Building and Environment*, 37: 1279–1288.

Shuttleworth, W.J. (1989). "Micrometeorology of tropical and temperate forest," *Philosophical Transactions of the Royal Society of London, B*324: 299–334.

Simpson, J.R. and E.G. McPherson (1997). "The effects of roof albedo modification on cooling loads of scale model residences in Tucson, Arizona," *Energy and Buildings*, 25(2): 127–137.

Simpson, J.R. and E.G. McPherson (1998). "Simulation of tree shade impacts on residential energy use for space conditioning in Sacramento," *Atmospheric Environment*, 32(1): 69–74.

Souch, C.A. and C. Souch (1993). "The effect of trees on summertime below canopy urban climates: a case study Bloomington, Indiana," *Journal of Arboriculture*, 19(5): 303–312.

Stark, T.F. and D.R. Miller (1977). "Effect of synthetic surfaces and vegetation in urban areas on human energy balance and comfort,"

Proceedings of a Conference on Metropolitan Physical Environment, Syracuse, N.Y., 25–29 August 1975. USDA Forest Service Technical Report NE-25, Upper Darby, Pa.: Northeastern Forest Experiment Station, pp. 139–151.

Steyn, D.G. (1980). "The calculation of view factors from fisheye lens photographs," *Atmosphere-Ocean*, 18: 254.

Sundborg, A. (1950). "Local climatological studies of the temperature conditions in an urban area," *Tellus*, 2: 221–231.

Swaid, H. (1992). "Intelligent urban form (IUF): a new climate-concerned urban planning strategy," *Theoretical and Applied Climatology*, 46(2–3): 179–191.

Swaid, H. (1993a). "Urban climate effects of artificial heat sources and ground shadowing by building," *International Journal of Climatology*, 13: 797–812.

Swaid, H. (1993b). "Numerical investigation into the influence of geometry and construction materials on urban street climate," *Physical Geography*, 14(4): 342–358.

Swaid, H. and M.E. Hoffman (1990a) "Prediction of urban air temperature variations using the analytical CTTC model," *Energy and Buildings*, 14: 313–324.

Swaid, H. and M.E. Hoffman (1990b) "Climatic impacts of urban design features for high- and mid-latitude cities," *Energy and Buildings*, 14: 325–336.

Swaid, H., M. Bar-El, and M.E. Hoffman (1993). "A bioclimatic design methodology for urban outdoor spaces," *Theoretical and Applied Climatology*, 48: 49–61.

Taha, H., H. Akbari, A. Rosenfeld, and J. Huang (1988). "Residential cooling loads and the urban albedo – the effects of albedo," *Building and Environment*, 23(4): 271–283.

Taha, H., S. Douglas, and J. Haney (1997). "Mesoscale meteorological and air quality impacts of increased urban albedo and vegetation," *Energy and Buildings*, 25(2): 169–177.

Takakura, T., S. Kitade, and E. Goto (2000). "Cooling effect of greenery cover over a building," *Energy and Buildings*, 31: 1–6.

Tanikawa, M. (2002). "Tokyoites feel the heat: foreign birds flourish as the city faces crisis," *International Herald Tribune*, August 29.

Terjung, W.H. and S.S.-F. Louie (1973). "Solar radiation and urban heat islands," *Annals of the Association of American Geographers*, 63: 181–207.

Terjung, W.H. and S.S.-F. Louie (1974). "A climatic model of urban energy budgets," *Geographic Analysis*, 6: 341–367.

Todhunter, P.E. (1990a). "Microclimatic variations attributable to urban-canyon asymmetry and orientation," *Physical Geography*, 11(2): 131–141.

Todhunter, P.E. (1990b). "The response of urban canyon energy budgets to variable synoptic weather types – a simulation approach," *Atmospheric Environment*, 24B(1): 35–42.

Tso, C.P. (1996). "A survey of urban heat island studies in two tropical cities," *Atmospheric Environment*, 30(3): 507–519.

Tuller, S.E. (1975). "The energy budget of man: variations with aspect in a downtown urban environment," *International Journal of Biometeorology*, 19(1): 2–13.

Ulrich, R.S. (1984). "View through a window may influence recovery from surgery," *Science*, 224: 420–421, (27 April).

Vukovich, F.M., J.W. Dunn III, and B.W. Crissman (1976). "A theoretical study of the St. Louis heat island: the wind and temperature distribution," *Journal of Applied Meteorology*, 15(5): 417–440.

Watson, I.D. and G.T. Johnson (1987). "Graphical estimation of sky view factors in urban environments," *Journal of Climatology*, 7: 193–197.

Wanner, H. and J.-A. Hertig (1984). "Studies of urban climates and air pollution in Switzerland," *Journal of Climate and Applied Meteorology*, 23(12): 1614–1625.

Webster, N. (1799). *A Brief History of Epidemic and Pestilential Diseases*, 2 vols, Hartford, Conn.: Hudson and Goodwin.

Whitford, V., A.R. Ennos, and J.F. Handley (2001). "'City form and natural process' – indicators for the ecological performance of urban areas and their application to Merseyside, UK," *Landscape and Urban Planning*, 57: 91–103.

Wienert, U. and W. Kuttler (2001). "Statistical analysis of the dependence of urban heat island intensity on latitude," *Proceedings of the 4th Symposium on Urban Environment*, American Meteorological Society, Paper No. 5.8, http://ams.confex.com/ams/AFMAPULE/4Urban/program.htm.

Willeke, D.C. (1989). "The imperative forest," in G. Moll, and S. Ebenreck (eds), *Shading our Cities: A Resource Guide for Urban and Community Forests*, Washington, D.C.: Island Press, pp. 58–63.

Wong, N.H., S.F. Tay, R. Wong, C.L. Ong, and A. Sia (2002). "Life cycle cost analysis of rooftop gardens in Singapore," *Building and Environment*, 38(3): 499–509.

World Meteorological Organization (WMO) (1970). *Urban Climates*, Vol. I, WMO Technical Note 108, Geneva, Switzerland: WMO.

Yamashita, S., K. Sekine, M. Shoda, K. Yamashita, and Y. Hara (1986). "On relationships between heat island and sky view factor in the cities of Tama river basin, Japan," *Atmospheric Environment*, 20(4): 681–686.

Thermal comfort in the urban tropics

CHAPTER 3

We have now looked at the urban climate anomaly, its primary causes and the energetic basis of the phenomenon. While these facets of urban climate indicate the context in which climate-sensitive design must occur, it is also necessary to focus on the goals of design in the altered urban context. The goal of climate sensitive design in any context is captured by the term "bio-climatic design."

Simply put, bio-climatic design is about designing for the human being. A term coined in the early 1960s by the Olgyay brothers (see Olgyay and Olgyay, 1963), bio-climatic design attempts to fashion architecture in harmony with nature while keeping the comfort needs of the human being as its central concern. The connection between the principles of sustainable design and bio-climatic design is hard to miss (Roulet, 2001). In an indirect manner, the principles of bio-climatic design are also enshrined in Agenda 21 as an important goal in achieving sustainable development.

The principles of sustainable design encompass respect for the environment, frugality in the use of energy and material, and a holistic outlook. However, bio-climatic design, with its central concern for the well-being of *humans*, adds a crucial dimension to sustainable design by reminding us of the need to fulfill a deeply felt human need.

It is sometimes argued that concern for nature should take precedence over concern for the human being. However, the principles of sustainability are truly fulfilled only if human comfort considerations are also met. While the savings in energy and construction materials and the protection of the environment are worthy causes in their own rights, these will not be sustainable unless the human needs are met (see Roulet, 2001). The question, then, is not how to save energy in buildings but how to achieve energy savings without sacrificing human comfort needs. A bio-climatic approach to architectural design offers a possible solution to this central question.

Although the term "bio-climatic design" is new to the architectural world, ancient builders were aware of the need for human-centered climatic design. Traditional building design took the climatic "given" as the starting point and derived not only building forms and practices but also generated cultural attributes. Over centuries, this trial-and-error evolution was able to produce "traditional" design solutions that are climatically appropriate, culturally relevant and aesthetically pleasing.

Unfortunately modern societies seem to have forgotten this art. And this failure to respect the climate and design with the climate is most evident in the tropics. The ubiquitous steel and glass box where curtain walls and claddings "seal-in" the indoor from the external conditions in the tropics goes against every known principle of sustainable design.

We begin our discussions by looking at the bio-climatic needs of humans in general and in the tropical context in particular. We then look at the issues related to the quantification of tropical thermal comfort, particularly the confounding effects of acclimatization, adaptation and expectation. Having established the thermal comfort patterns in the tropics, we turn our attention to bio-climatic requirements specific to tropical cities, and outline the role of shading, ventilation and vegetation in enhancing thermal comfort in the urban tropics.

3.1 Bio-climatic needs of humans

Human deep body temperature (the so-called "core temperature") must be maintained at about 98.6°F (37°C) for health. Since metabolized food releases energy, a healthy human body normally attempts to *lose* heat to the ambient environment at all times (Figures 3.1). The body can – for short duration – gain heat, although this is not desirable for longer periods of time.

Under moderate environmental conditions, the human body immersed in the environment for at least an hour strives to achieve thermal balance. This steady-state heat flow per unit area per unit time is given by Fanger (1970: 22–23):

$$H - E_d - E_{sw} - E_{re} - L = K = R + C, \qquad (1)$$

Where H = internal heat production in the human body
E_d = heat loss by water vapor diffusion through the skin
E_{sw} = heat loss by sweat evaporation
E_{re} = sensible heat loss by evaporation
L = latent heat loss by respiration
K = conduction from outer surface of clothed body
R = radiation loss from outer surface of clothed body
C = convection heat loss from outer surface

H is a function of the level of human activity. The higher the activity the more the heat loss has to be for thermal balance. E_d is a function of the difference between saturated vapor pressure at skin temperature and the actual vapor pressure in ambient air (Fanger, 1970). Since air temperature and humidity are high in the tropics, heat loss due to E_d is negligible.

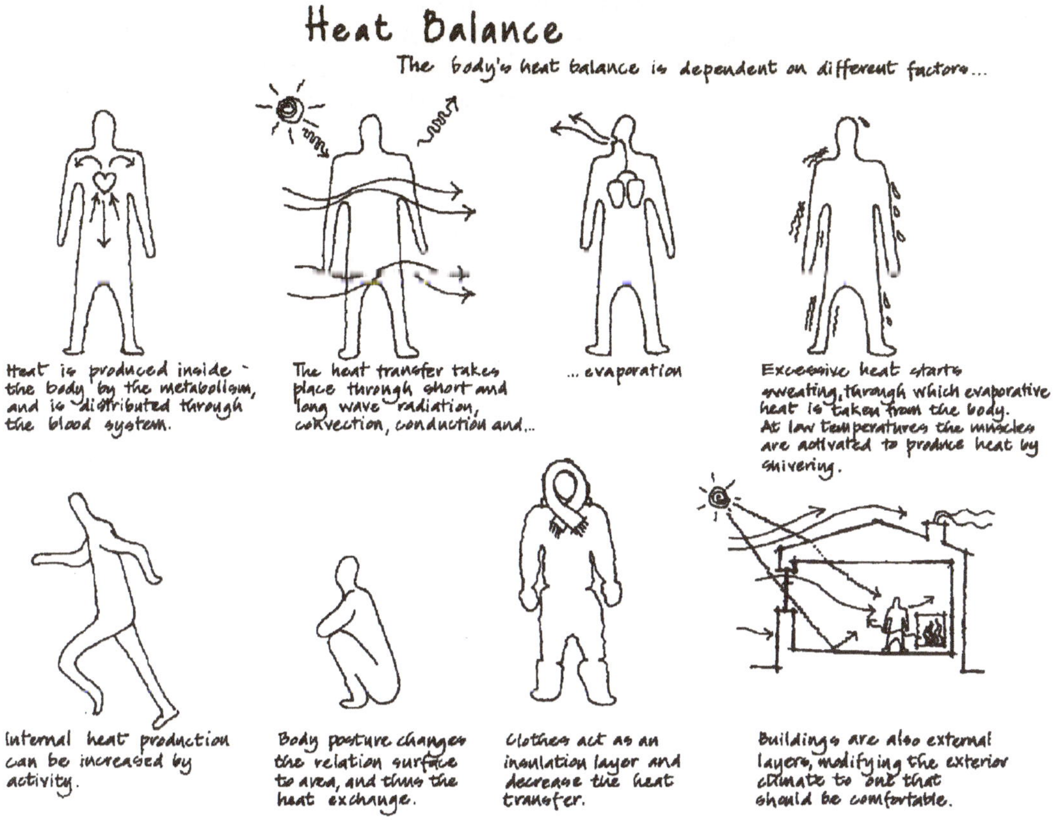

3.1 *Human energy balance: steady-state conditions*

L is a function of the humidity ratio difference between the inhaled and the exhaled air. Here, too, high humidity in the hot-humid region dictates that *L* will be low. E_{re} is usually low for all environments (Fanger, 1970: 29). Unless the body is in a reclined position, conduction loss (*K*) is usually negligible too.

At "comfortable" air temperatures (68–77°F or 20–25°C), and with little wind movement (<0.33 ft s^{-1} or 0.22 m.p.h. or 0.1 m s^{-1}), the metabolic heat production is mainly dissipated by radiation (*R*), accounting for up to 60 percent of heat loss. Convection loss at this stage will account for 15 percent of the heat produced, while evaporation from the lungs and through the skin (E_{sw}) accounts for about 25 percent heat loss (Oke, 1987). However, an increase in wind flow will proportionately increase evaporative and sensible heat losses.

At warmer environments (air temperature >80°F or 26.6°C), the human thermoregulatory mechanism cannot depend on radiation, for radiation is a function of the temperature *difference* between the source and the sink (in this case, between a human body and its surrounding environment). Thermoregulation by convection (*C*) becomes insignificant if ambient air temperature rises above 95°F

(35°C). At this point, the air is warm enough to act as a heat *source* to the body rather than as a sink (Oke, 1987: 223). Therefore, the body dilutes the blood vessels just under the skin in an attempt to reduce the insulation potential of the skin (thus the skin becomes red with exposure to the warm environment) (Fanger, 1970). By diluting the blood vessels, more warm blood flows very near the surface of the skin, and is able to lose some of its heat to the outside environment via radiation and/or convection. However, when the air temperature crosses the 95°F (35°C) threshold, even this mode of heat loss becomes insignificant. Therefore, at higher ambient air temperatures heat loss by humans must depend on evaporation (E_{sw}).

Monteith (1973) mentions that a healthy adult male can lose up to 1 kg of water every hour. This loss translates into an hourly energy loss of 375 W/m^2. Although such large quantities of heat loss should be sufficient to cool the body in warmer environments, the actual evaporation in human beings is somewhat less due to the increasing presence of salt on the skin deposited by evaporating sweat. Prolonged evaporative heat losses of higher rates are lethal. When an average man loses 2 percent of his body weight equivalent of water (about 0.72 kg or 1.6 lb) by evaporation, he becomes thirsty. At 4 percent loss he feels apathetic and impatient, at 8 percent speech becomes difficult and when the water loss reaches 18–20 percent of body weight equivalent death is imminent (Oke, 1987: 225).

Culturally, therefore, tropical dwellers have attempted to enhance the thermoregulatory processes mainly through built envelope manipulations, clothing modifications and maintaining a rhythmic balance of activities in conjunction with the climatic seasons. Lightly clad, open form buildings are the traditional norm in these societies. In terms of clothing, the custom is to utilize loosely woven, thin materials covering a bare minimum of body, worn with minimal underclothes (Mather, 1974). In the absence of strongly differentiated seasons, activity cycles in the tropics are more diurnal than seasonal. Traditional societies in the tropics have developed working habits beginning earlier in the day when the sun is low, or late in the evening soon after sunset (Correa, 1989). These actions, which we will later call adaptive behavior, enable tropical residents to tolerate a wide range of ambient conditions through the manipulation of the cultural environment as well as the thermoregulatory processes.

3.2 Thermal comfort in the tropics

While tropical human thermoregulatory processes are similar to those in other areas, tropical climate poses peculiar comfort challenges. The tropical thermal comfort challenge is well captured by Figure 3.2, which indicates the monthly maximum and minimum dry bulb temperatures and relative humidities for a tropical city (Colombo, Sri Lanka) on

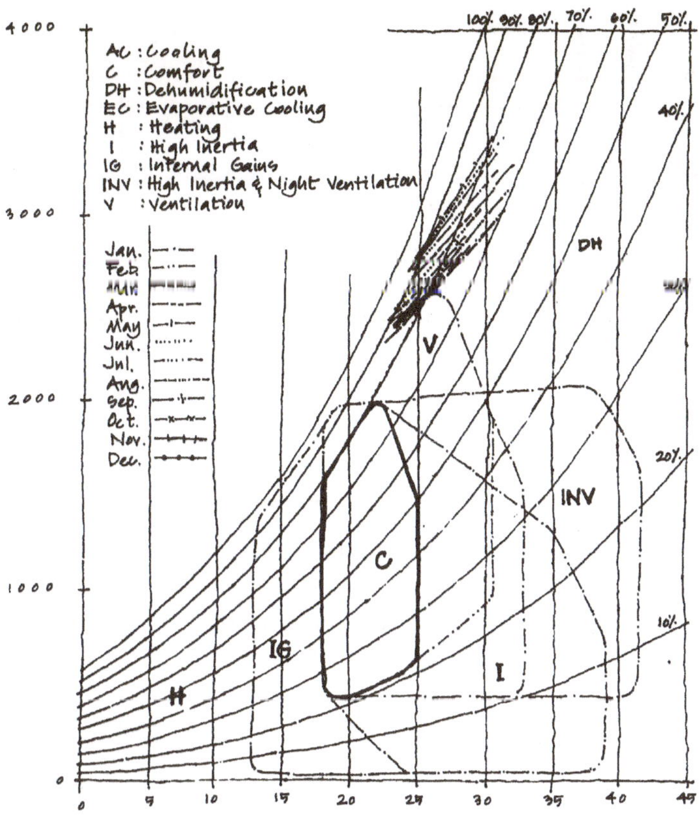

3.2 *Typical thermal comfort conditions in the tropics: Colombo*

the ASHRAE's psychrometric chart. It is clear that tropical conditions are above the comfort zone and require design and building fabric devices to achieve comfort (see Khedari *et al.*, 2000). At such high temperatures and humidity, the primary – if not the only – heat balance mechanism available to an average person is evaporation (Figure 3.3). And the primary mode of facilitating evaporation is air movement.

The cooling effect of air movement depends not only on the speed of air but also on temperature, humidity and radiation balance as well as the activity and clothing level of an individual. Perceptible air movement encourages sweat evaporation, which in turn takes heat away from the skin: a cooling sensation therefore results. This in turn permits an individual to "tolerate" higher air temperatures that would normally be considered oppressive in the absence of significant air movement. In other words, air movement appears to extend the limits of "acceptable temperatures" by making higher temperatures feel cooler than they really are.

The precise amount of apparent reduction in air temperature due to air movement (the so-called "cooling effect of air movement") has been the subject of considerable speculation for over seventy years.

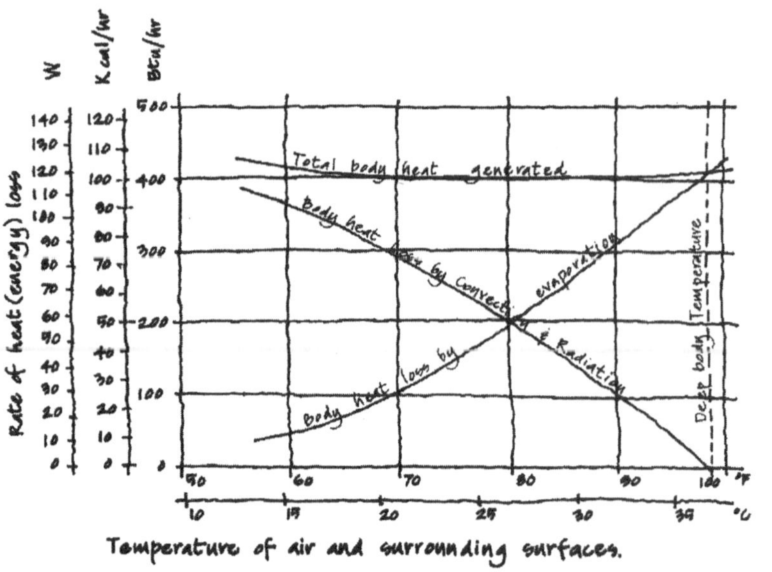

3.3 *Typical body-heat loss for a lightly clad male subject*

Szokolay (2000) classified major efforts at quantification of the cooling effect of air movement into two categories:

1. those that can be expressed by a linear function with, in most cases, an upper limit;
2. those that recognize the law of diminishing returns (i.e. increases in cooling effect will reduce with elevated air velocities).

Table 3.1 lists the major recommendations under both categories. In the table, dT indicates the cooling effect in °K (i.e. the apparent reduction in air temperature) and v denotes air movement at body level (in m/s). The relationships are derived from Szokolay (2000: 146ff.).

Figure 3.4 graphically represents the recommended cooling effects, and a compromise function developed by Szokolay (2000) is given in equation (2).

$$dT = 6 \times v_e - 1.6 \times v_e^2 \qquad (2)$$

Where, v_e is the effective velocity (in m/s). Effective velocity is arrived at by deducting 0.2 m/s from body level air movement ($v_e = v - 0.2$).

Thermal comfort indices for the tropics

Thermal comfort quantification can take one of two forms:

- field studies;
- rational approach (see Nicol and Humphreys, 2002).

Table 3.1 Recommendations of the cooling effect of air movement

Author	*Cooling effect*	
	Function	*Calculated values*
Linear functions		
Drysdale (1952)	$dT = 6.7*v$	
Arens and Watanabe (1986)	$dT = 8.27*(v-0.25)$	
ASHRAE (1985)	$dT = 3.6*(v-0.25)$	
Rohles *et al.* (1974)	$dT = 2.08*v$	
Asymptotic functions		
Humphreys and Nicol (1995)	$dT = 7 - \left(\frac{50}{4 + 10\sqrt{v}}\right)$	
Arens *et al.* (1981)		0.5 m/s → 2.0 K
– at 50 percent RH		1.0 m/s → 4.1 K
		2.0 m/s → 5.4 K
– at 30 percent RH		0.5 m/s → 2.2 K
		1.0 m/s → 4.5 K
		2.0 m/s → 6.5 K
– at 12 g/kg abs. hum.		0.5 m/s → 2.5 K
		1.0 m/s → 5.0 K
		2.0 m/s → 7.5 K
ASHVE (1932)		0.3 m/s → 0.8 K
		0.5 m/s → 1.6 K
		1.0 m/s → 2.9 K
		1.5 m/s → 3.7 K
		2.0 m/s → 4.0 K
ASHRAE (1992)		Up to 0.8 m/s → 3.0 K
		1.0 m/s → >3.0 K
		up to 1.5 m/s → 3.7 K

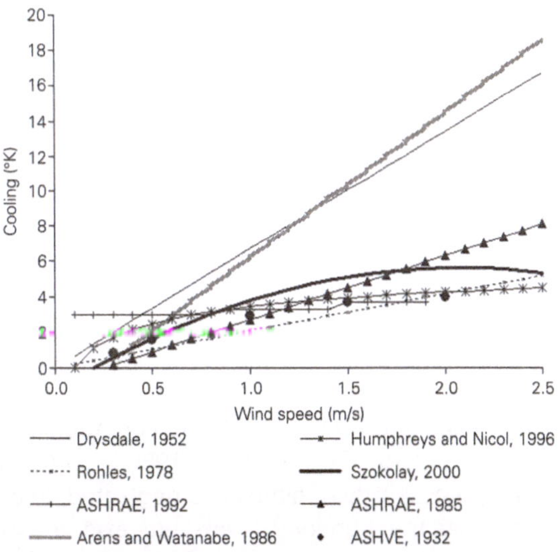

3.4 *Cooling effect of air movement*

The former is based on findings of surveys of thermal comfort conducted in the field. The field study approach gathers field survey data on the thermal environment and the subjects' thermal sensation, and attempts to correlate both statistically. In the latter approach, a human energy balance model (such as the one presented in the first section of this chapter) is used to explain the response of people to the thermal environment in terms of the physics and physiology of heat transfer. An "index" is thereafter developed to express the thermal state of the human body based on controlled experiments in test chambers.

Field survey approach to quantifying thermal comfort in the tropics began in the 1940s (Rivera de Figueroa, 1980). These include Ellis (1952, 1953), who recorded "comfort" (or neutral) temperature to be 26.1–26.7°C (79–80°F) in Singapore; Webb (1959), recording a comfort temperature of 27.2°C ((81°F) also in Singapore; Rao (1952) in Calcutta (26.0°C [78.8°F]); and Nicol (1974), recording 31.1°C (88°F) in Roorkee, India. Additionally, many qualitative (descriptive) studies on thermal comfort in the hot-humid regions were also carried out. These include Ambler (1955), Ballantyne *et al.* (1967) in Port Moresby, Papua New Guinea, and Wyndham (1963) in Tropical Australia.

a) Effective temperature (ET) is defined as that temperature at 50 percent relative humidity and with mean radiant temperature equal to air temperature which produces the same thermal sensation as the actual environment; that is, ET normalizes temperatures for humidity and radiation and thereby enables comparisons of various thermal environments (Yaglou and Miller, 1924).

Disadvantages in using ET for the quantification of tropical thermal comfort are:

- the cooling potential of air movement is not accounted for;
- the insulation effects of clothing are disregarded;
- heat loss requirements varying with human metabolism are ignored.

(b) Standard effective temperature (SET) was developed by the American Society of Heating, Refrigerating and Air Conditioning Engineers (ASHRAE) in order to rectify some of the shortcomings of ET. SET is defined as the equivalent dry-bulb temperature of an isothermal environment at 50 percent relative humidity in which a subject, while wearing clothing standardized for the activity concerned, would have the same heat stress and thermoregulatory strain (skin wetness) as in the actual test environment after an hour's exposure to it (ASHRAE, 1985).

Although SET takes into account all the parameters of comfort identified by Fanger (1970), Gagge *et al.* (1986) point out that the effect of the vapor permeability of clothing is neglected. In a hot-humid environment, not only the humidity levels but also the porosity of clothing will determine how much sweat *actually* evaporates. If the clothing worn is of low vapor permeability, the maximum potential of evaporative cooling will not be available to the body.

There are many problems with field studies in the tropics. First, typically only two environmental variables are measured (usually, air temperature and humidity). Second, no account of clothing and activity levels of subjects are given. Third, the sample sizes are small (less than 20 in most cases). None of the indices based upon these "comfort" temperatures withstood the test of time. Only one of these indices, the "Tropical Comfort Index" (Webb, 1952/53) developed an applicable thermal comfort quantifying mechanism, but this too did not specify clothing levels. Furthermore the sample size was too small (20 mostly European subjects living in Singapore) to be statistically significant.

Current thermal comfort practice in the tropics therefore employs the more universally accepted "rational approach," even though these did not originate in the tropical region. We therefore discuss below some of the most commonly used rational indices developed in the temperate zone and point out their shortcomings in quantifying thermal comfort in the tropics.

(a) Predicted mean vote (PMV) by Fanger (1970) (also adapted by the International Standards Organization, ISO 7730, 1994) is defined as the heat load that would be required to restore a state of "comfort." "Comfort" here is defined in terms of four environmental parameters (air temperature, mean radiant temperature, relative humidity, and air velocity) and two cultural parameters (clothing and activity level). Comfort is measured on a seven-point subjective PMV scale, starting at –3 (cold), –2 (cool), –1 (slightly cool), 0 (neutral/comfortable), 1 (slightly warm), 2 (warm), and 3 (hot).

The PMV model is based on sweat secretion and mean skin temperature of the body. It predicts that at neutral temperature, 80 percent of the people will vote between ±1 on the PMV scale. Neutral temperature is assumed to be associated with imperceptible rates of sweating.

Table 3.2 Limitations to the range of conditions over which PMV is applicable

Variable	*Units*	*Lower limit*	*Upper limit*
Metabolic rate	W/m^2 (met)	46 (0.8)	232 (4.0)
Clothing insulation	°C/W (clo)	0 (0)	0.31 (2.0)
Air temperature	°C	10	30
Radiant temperature	°C	10	40
Air velocity	m/s	0	1.0
Vapor pressure	P_a		
Predicted Mean Vote	PMV	–2	+2

While PMV takes all the known comfort parameters into account, it assumes that comfort occurs when there is no perceptible sweating on the skin. It thus underestimates the potential for comfort even in the presence of sweat under naturally occurring environments in the hot-humid zone. Furthermore, it too fails to account for the vapor permeability of clothing.

(b) Winslow's skin wettedness index (DISC) is the fraction of body surface wet with perspiration required to regulate body temperature by evaporative cooling (Gagge *et al.*, 1986). In a sense the DISC index combines the shortcomings of PMV and SET.

(c) ANSI/ASHRAE 55-1992 (also adapted by the ISO, EN ISO 7730, 1994). This amended thermal comfort index superseded ANSI/ASHRAE 55-1981 which was based on the PMV model of thermal comfort (Fanger, 1970). In the revised standard, insulation values of clothing have been revised so as to reflect the correct *ensemble* clothing insulation (ASHRAE, 1992). Also, the predicted mean vote scale now has nine points instead of the original seven-point original PMV scale (the extra points being ±4, very hot/very cold respectively).

Problems with using rational thermal comfort indices in the tropics

Subsequent to the adaptation by ASHRAE in 1981 (and the modification in 1992), the PMV-based ANSI/ASHRAE 55-1992 (also adopted by the ISO: ISO 7730, 1994) has become the most commonly used comfort index in the field of human thermal comfort all over the world (de Dear and Fountain, 1994). With rapid economic growth in the hot-humid areas, particularly the tropical regions of Asia and Latin America, demand for HVAC-based space conditioning is on the rise. This section therefore looks at the shortcomings in using the PMV-based ASHRAE comfort index for the tropics.

It is becoming increasingly clear that the comfort predictions made by the ASHRAE/ISO models are vastly different to the lived experience of tropical dwellers. Why are predictions based on rational approach to thermal comfort wrong, particularly in the tropics?

The basic PMV model of thermal comfort was developed by Fanger (1970) using college-aged American and Danish students responding to thermal environments in climate chambers. The debate over the relevance of results from such controlled environments to the "real" world has raged ever since it was adapted as a universal standard (see, for example, Rohles, 1978; McIntyre, 1982; Prins, 1992).

There are at least three inherent limitations to the rational approach, especially the PMV approach defined in the ASHRAE's current standard (ANSI/ASHRAE 55-1992) and the ISO model (EN ISO 7730, 1994):

- limitations of applicability;
- basic assumption of the heat balance approach;
- problems with quantifying clothing insulation and/or metabolic rate (see Nicol, unpub.)

The main problem for the prediction of comfort using ISO 7730 in tropical climates is the declared limitations of applicability of the PMV model (see Table 3.2). Air temperatures above 30°C and velocities in excess of 1 m/s are very common in the tropical building, especially if they are naturally ventilated. With the use of fans, etc., many tropical dwellers find these conditions comfortable, while the PMV model would predict comfort sensation to be warm to hot.

The second problem with rational models is that the heat balance approach to thermal comfort (i.e. heat loss from the body equals metabolic heat production of the body) assumes that thermal sensation will not be neutral if the body is not in equilibrium. However, this formulation is problematic in that it is *not* possible for an imbalance to persist in the heat flow between the body and the environment: death would result (see Humphreys and Nicol, 1996). In other words, *a persistent positive or negative PMV is theoretically impossible.* If people are uncomfortable, they will take actions to achieve comfort, such as changing their clothing, activity levels or, where possible, opening windows, closing the curtains/blinds, or switching on a fan, etc.

Thirdly, quantification of clothing and metabolic activity is also problematic. Heat balance approach sees clothing as an insulator, but in the tropics clothing could also be used to maintain an appropriate microclimate next to the skin (Berger, 1988, quoted by Nicol, unpub.). Additionally, the heat balance approach assumes metabolic rate to be independent of the external environment. This is patently wrong: people adjust their activity levels, as well as posture, to affect metabolic rate in a changing external environment.

Another criticism stems from the semantic differences between neutrality, acceptability and preference (Busch, 1992). While the PMV model is expected to predict the "neutral" temperature (i.e. temperature at which no perceptible sweating occurs), it does so by asking people to vote for their climatic environment. What is being voted could be both "acceptable" as well as "preferable." While an environment may be acceptable to a person, he/she might *prefer* a better environment. This difference is lost to the PMV model:

> Each of these effects may be a minor source of error in itself, but thermal behavior is essentially governed by the thermal sensation in a feedback relationship where the sensation is the 'trigger' for a change in the environment. Thus a number of measures may be taken and they are often additive so small errors in estimating each can result in significant errors in the overall prediction. A model which represents the thermal equilibrium as a heat balance at a

> point in time will not fully reflect thermal comfort in the [real world]. Without full allowance for the dynamic nature of the human interaction with their surroundings, such a model is likely to have limited applicability. This is especially true in conditions found in free-running buildings in tropical climates (Nicol, unpub.).

In order to test the validity of the PMV approach to thermal comfort, ASHRAE began funding a series of field studies of thermal comfort in the mid-1980s in office buildings spread across four different climate zones. As part of this effort, ASHRAE funded a research project that collected 21,000 sets of standardized thermal comfort data from 160 different buildings located in Australia, Canada, Greece, Indonesia, Pakistan, Singapore, Thailand, UK and the USA (de Dear *et al.*, 1998). A remarkable finding of the meta-analysis using this high-quality database from a variety of building types and diverse cultural backgrounds is that the PMV model is unbiased when used to predict comfort temperature in air-conditioned buildings. However, the PMV model overestimates the subjective warmth sensation in naturally ventilated buildings in warm climates (de Dear and Brager, 1998). This is clear from Figure 3.5, where even with a large variation in outdoor temperature (from −24 to +34°C), the indoor climate preferred by people in climate-controlled buildings varied from 17–23°C only. The situation is vastly different for non-AC buildings. By extension, we can conclude that in the non-climate controlled urban outdoors the rational approach to thermal comfort will simply overestimate the thermal comfort problem.

Adaptation, acclimatization and thermal expectation – a tropical perspective

The preceding section outlined the mismatch between the comfort conditions predicted by the "rational approach" and the everyday experience of tropical dwellers. The disparity between the two has led

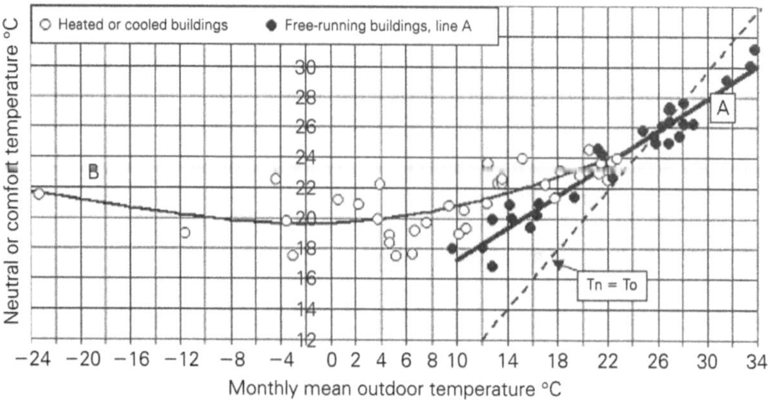

3.5 *Relationship between "comfort" temperature and DBT.* Source: Nicol (unpub.)

to a host of problems, ranging from excessive wastage of energy to sick building syndrome (SBS). We have now looked at the systemic reasons for this mismatch, but a more qualitative description of the underlying causes is also necessary if we are to quantify thermal comfort in the tropical urban outdoors realistically.

Two of the most plausible qualitative reasons for the mismatch between comfort predictions of the rational approach and the lived experience of tropical dwellers are succinctly portrayed by the concepts of *adaptive approach* and *thermal expectation*. Variously known as adaptation and acclimatization, the fundamental assumption of the adaptive approach is expressed thus: "*If a change occurs such as to produce discomfort, people react in ways which tend to restore their comfort*" (Nicol and Humphreys, 2002: 564).

By linking the comfort vote to people's actions, the adaptive principle links the comfort temperature to the context in which people find themselves. The comfort temperature is a result of the interaction between people and the environment they occupy. The options for people to react will reflect their situation: those with more opportunities to adapt themselves to the environment or the environment to their own requirements will be less likely to suffer thermal discomfort (Nicol and Humphreys, 2002: 564).

In an extensive survey of thermal sensation and the thermal comfort of apartment dwellers in Singapore's high-density housing estates, Wong *et al.* (2002) found that even though people generally experienced warmer thermal sensations (i.e. +2 and +3 on the ASHRAE scale, "warm" to "hot"), they still found conditions to be acceptable. This was because of the opportunity they had to engage in adaptive behavior. The most common adaptive behaviors exhibited by apartment dwellers in Singapore were (in descending order of importance): fans, opening of windows, drawing of curtains/blinds (environmental adaptation), drinking more water, adjusting clothing, and taking frequent baths (personal adaptation).

People's ability to adapt depends on a number of factors. Prominent among them are indoor climate, outdoor climate and type of buildings. The last principle is more attuned to the notion of "expectation" and is discussed on pp. 78–81. Our attention here is drawn to the prime contextual variable, which is *climate*.

Evidence has been mounting for over thirty years that comfort temperature is a function of climate, particularly outdoor climate. Auliciems (1983) and Humphreys (1981), having analyzed over fifty thermal comfort studies utilizing over 250,000 subjects, found strong correlation between the reported comfort temperatures and the outdoor temperatures under which they were obtained.

However, the strongest evidence for adaptation to the outside conditions has come from the hot-humid areas. Four studies are reported here to show how people's previous experience with a particular climate influences their present expectations of thermal comfort.

Basnayake (1984) conducted a time-series analysis of thermal comfort variation in Sri Lanka to ascertain whether comfort temperatures changed due to seasonal and diurnal variations in climate. It was reported that comfort temperatures rose in the warmer season and fell in the less warm season. Similar patterns were detected in a diurnal cycle as well. Furthermore, a seasonal acclimatizational effect was detected in which the change in ambient temperature which produced thermal discomfort was found to be larger in the warm months than in the cool months (Basnayake, 1984: 191).

Busch (1992) studied two groups of office workers in Bangkok, one working in air-conditioned offices and the other in naturally ventilated offices. He found that the upper bound for comfort air temperatures preferred by workers in air-conditioned offices was 28°C (83°F) while that of the workers in naturally ventilated offices was 31°C (88°F). Both were significantly higher than the current standard for summer-time thermal comfort (ANSI/ASHRAE 55-1992), which is 22.8–26.1°C (73–79°F). This study also showed that workers in the air-conditioned offices wore clothing that had, on average, 15 percent more insulation value than those worn by workers in the naturally ventilated offices. Furthermore, the variability of clothing insulation among naturally ventilated office workers was much smaller than that of the air-conditioned office workers (0.48 clo. and 0.95 clo. units respectively). This indicates that the workers in the air-conditioned offices either got used to the cooler environment or have come to *expect* such coolness that they wear clothing with higher insulation. Busch concluded that "while Thai office workers from different thermal environments may be similarly sensitive to thermal change, the thermal levels to which they respond are different" (Busch, 1992: 245).

De Dear *et al.* (1991a) conducted a similar study in Singapore. They too found that thermal requirements inside tropical buildings, particularly those that are naturally ventilated, could be significantly warmer (about 3.0°C [5.4°F] than the air-conditioned ones) and more acceptable to their occupants than those predicted by the PMV model.

De Dear and Fountain (1994) conducted a survey of thermal comfort among office workers accustomed to be in air-conditioned offices and homes in Townsville, Queensland, Australia. Many of the workers in this tropical city (latitude = 19°S) have grown accustomed to living in air-conditioned homes, and their thermal comfort preferences (at 23.5°C or 74.3°F) were found to be almost exactly as those predicted by the PMV model for those conditions. They concluded that these residents, having been accustomed to air-conditioned living, now expect low-comfort temperatures – and that is what they preferred as well.

In the last three studies, the PMV models predicted the neutrality temperatures well. For example, the Townsville office workers voted their environment to be –0.4 on the PMV scale (between comfortable and slightly cool) during the dry period, the comfort model predicted the vote to be –0.2. During the wet period, both observed and predicted comfort votes were –0.3. Furthermore, climate chamber studies in hot-humid Singapore (de Dear *et al.*, 1991b) and Tokyo (Tanabe *et al.*, 1987) also found neutrality temperatures (i.e. air temperatures most frequently associated with a PMV vote of zero) not significantly different from the 25.6°C (78°F) neutrality found by Fanger (1970) using Danish and American college-aged students.

It is thus increasingly becoming clear that tropical thermal comfort, particularly in naturally ventilated environments, is a function of outdoor climate. These comfort temperatures do not resemble anything predicted by the PMV model. However, the neutral temperature predictions by the PMV model are still valid, even in the tropics.

Comparing studies done in the 1970s (Humphreys, 1978) with those from the 1990s (de Dear, 1998), we can state that the following relationship sufficiently explains the adaptive principle based on external climate (see Nicol and Humphreys, 2002):

$$T_c = 13.5 + 0.54T_o \qquad (3)$$

Where T_c is the comfort temperature and T_o is the average mean outdoor temperature. The comfort temperature derived from equation (3) represents the upper limit of what 90 percent of the people are likely to find comfortable.

In recent times, de Dear and Brager (2001, 2002) have attempted to develop what they term an "adaptive comfort standard" (ACS). The ASHRAE has been attempting to review its current thermal comfort standards with a view towards incorporating a "variable temperature standard" since 1998. This resulted in a public review draft of the ASHRAE Standard 55-1992 (amended in 1995) being issued in February 2001. Currently, the ACS is listed in the appendix to Standard 55.

The ACS presented below is shown as an optional method for determining acceptable thermal conditions in naturally conditioned spaces. Figure 3.6 shows the ACS derived from temperature preferences in naturally conditioned spaces as a function of outdoor temperature. Outdoor air temperature (T_o) is simply the arithmetic average of the mean monthly minimum and maximum daily air temperatures for the month in question:

Lower 80 percent acceptable limit:	$T_c = 14.3 + 0.31\,T_o$	(4)
Lower 90 percent acceptable limit:	$T_c = 15.3 + 0.31\,T_o$	(5)
Optimum comfort temperature:	$T_c = 17.8 + 0.31\,T_o$	(6)
Upper 90 percent acceptable limit:	$T_c = 20.3 + 0.31\,T_o$	(7)
Upper 80 percent acceptable limit:	$T_c = 21.3 + 0.31\,T_o$	(8)

These are likely to be true for mean outdoor air temperatures up to 33°C. The 80 percent and 90 percent acceptability limits were not derived from empirical acceptability but are based on the principles of

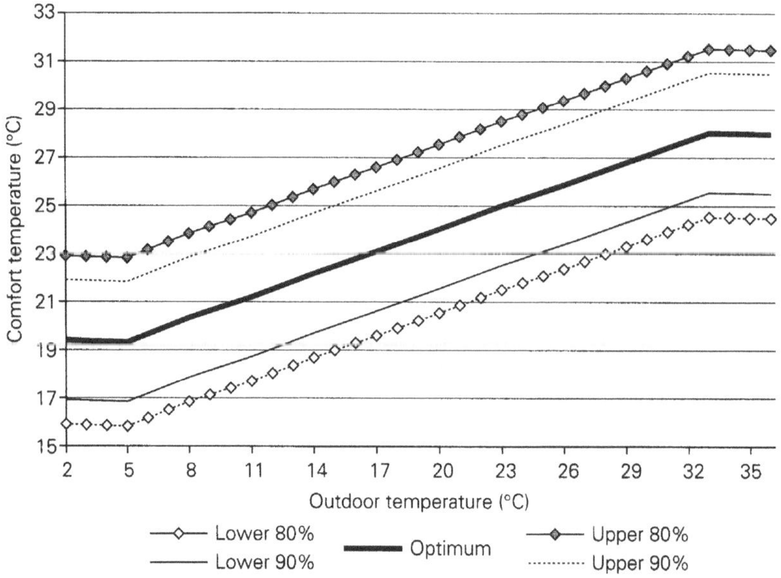

3.6 *The adaptive comfort standard (ACS) approach*

PMV (Fanger, 1970). However, the above equations are true for typical clothing levels normally expected from individuals adapted to the outdoor conditions indicated (i.e. average temperature range from 5–33°C). These functions cannot be extrapolated for temperatures outside the range indicated. Beyond 33°C (and below 5°C), the graph should flatten out, indicating constant comfort temperatures.

Thermal expectation

It is clear by now that thermal expectation is a crucial explanation for the difference between comfort model predictions and actual experience, particularly in the tropics (see Fanger and Toftum, 2002). People accustomed to non-air-conditioned buildings in the tropics are typically immersed in such environments for a very long time, perhaps even generations. "They may believe it is their destiny to live in environments where they feel warmer than neutral" (Fanger and Toftum, 2002: 534).

Analyzing the thermal comfort expectation, sensation and preference of Thai office workers moving from air-conditioned and non-air-conditioned buildings to outdoors and vice versa, Jitkhajornwanich and Pitts (2002) found that people working in air-conditioned buildings expected the outdoors to be warmer than it really was and voted accordingly. Similarly, even though they voted their sensation to be cool, people working in air-conditioned environments preferred even cooler temperatures on account of their expectation. This expectation is compounded by excessive dress code. Because air conditioning is available, people tend to over-dress (business suit in the tropics, for example). Furthermore, we can expect tropical dwellers to prefer cooler temperatures on account of their constant "expectation" of warm conditions.

The concept of "expectation" is now gaining wide currency. Recently, Fanger and Toftum (2002) have quantified thermal expectation in terms of an "expectancy factor" (*e*), which depends on the prevalence of air conditioners and the general warmness of the outdoor climate (Table 3.3). The expectancy factor is then used to adjust the predicted mean vote (PMV) sensation thus:

$$PMV_e = PMV \times e \qquad (9)$$

Where PMV_e is the predicted mean vote adjusted for expectation. It must however be stated that the "expectancy factor" proposed by Fanger and Toftum (2002) is still experimental and that independent verification of equation (9) has not been conducted. Furthermore, the suitability of the PMV approach itself, particularly in the tropical context, still remains an open question. In the stressful environment of the tropics, outdoor air temperature may ultimately prove to be the most easy-to-use predictor of comfort temperature.

In summary, we can state that the principles of adaptation, thermal expectation and preference, together with physiological processes, justify the thermal comfortability of a given situation (Figure 3.7). Thermal perception is not based on discrete human cognitive processes but rather on an accumulative process. If people are dissatisfied with a particular environmental condition they may either adapt themselves or the environment to achieve comfort or, in the absence of such possibilities, come to internalize a negative thermal expectation. If adaptation is possible, it may take one of two forms: personal and/or environmental adjustment.

However, the "neutrality" temperature is the same for all the groups studied, be they Danish or American college-aged students, Japanese schoolchildren or office workers in Bangkok, Singapore or Australia. The rational models appear to predict the neutral temperature in the hot-humid environment well, but not the preferred or acceptable temperature, which is a cultural condition. In a hot-humid environment people may even accept a warm environment by modifying their behavior, such as wearing lighter clothing, seeking

Table 3.3 Expectancy factors for non-air conditioned buildings in tropics

Expectation	*Classification of non air conditioned buildings*		*Expectancy factor*
	Prevalence of air-conditioning	*Length of warm period*	
High	Common	Brief (only during summer)	0.9–1.0
Moderate	Moderately common	All summer	0.7–0.9
Low	Rare	All year	0.5–0.7

Source: Fanger and Toftum (2002)

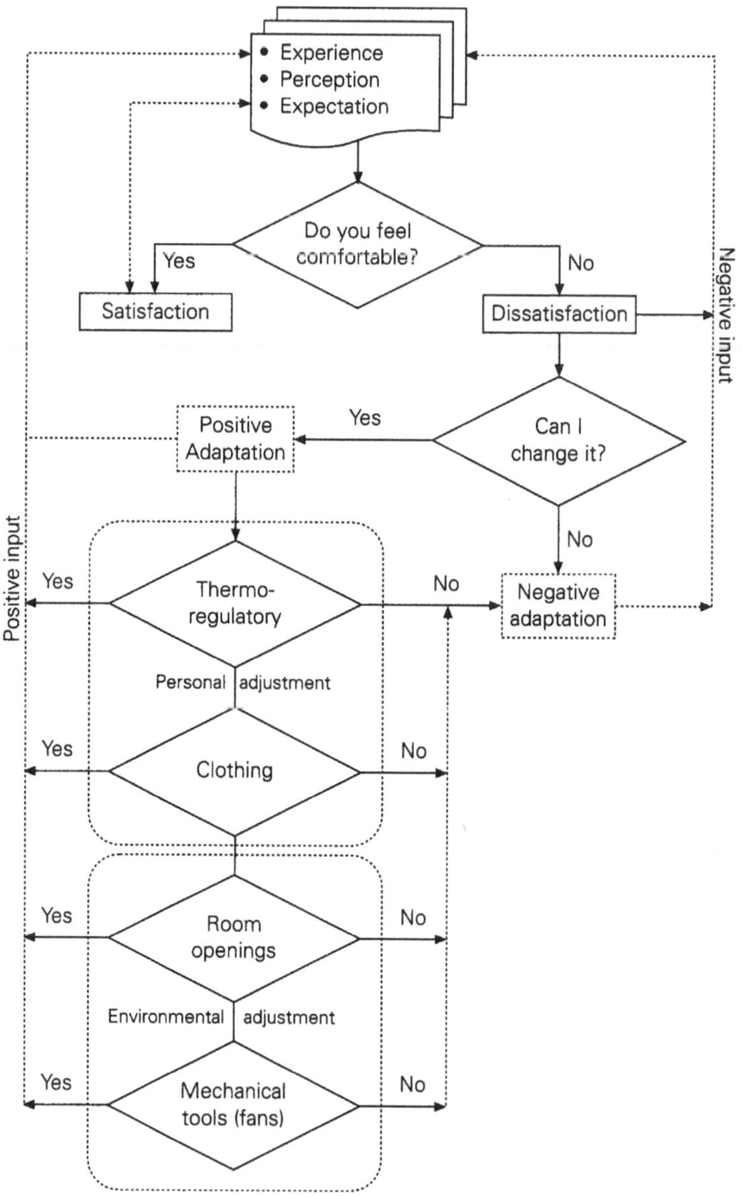

3.7 *Adaptive behavior and thermal comfort perception*

well-ventilated places, etc. In other words, people in the tropics seem to put their minds and bodies to adapt to heat (Prins, 1992).

What we know is this: the human body can survive in a wide range of thermal environments. This process is augmented by cultural artefacts like buildings, clothing and seasonal rituals. However, thermal comfort is achieved only within a very narrow range of climate change. At the same time the current thermal comfort standards based upon thermal chamber experiments on college-aged American and Danish students do not appear to be good indicators of thermal comfort in *actual practice*. It is increasingly becoming clear that this discrepancy is due to the human ability to adapt to the environments they have long been immersed in, even if these environments are somewhat extreme in terms of heat stress. By altering their behavioral patterns and habits, residents in the tropics are able to tolerate – indeed accept – much warmer temperatures than the comfort models have us believe.

The lack of fit between thermal comfort indices developed in the temperate areas and acceptable temperatures in the tropics highlights the need for empirical studies in the latter. In the face of rapid urbanization in the region the development of a more accurate and indigenous *urban thermal comfort index* assumes even greater urgency.

Having looked at thermal comfort standards extensively, we might posit the following question: Are thermal comfort standards (including the adaptive comfort standard) really necessary? Is it always desirable to have a fixed temperature in order to avoid discomfort at all times? Wouldn't it lead to thermal boredom? Kwok (2000) reviewed research and collected anecdotal evidence regarding the concept of thermal monotony and found that architectural educators typically encourage their students to explore and utilize the natural dynamic qualities of the thermal environment as inspiration for generating architectural form. Rather than aiming at thermal "neutrality" designers ought to strive for "pleasantness" where "thermal delight" (see Heschong, 1979) is possible. Although thermal comfort standards need not do this, it is important for architects to keep in mind that thermal comfort is not a static phenomenon, and our design goal should be to encourage as much diversity in our thermal environment as possible. What the ACS and other similar attempts seem to have achieved is to force much-needed scientific attention on these issues.

3.3 Thermal comfort in the tropical urban outdoors

The quantification of urban (or outdoor) thermal comfort is a relatively new area of inquiry. This is understandable considering the great complexity of issues involved in the urban environment: spatial and temporal variability of environmental conditions, diverse range of activities and clothing patterns, and complex effects of buildings and vegetation on shading and ventilation. Beginning in the 1930s, early efforts concentrated on the *outdoor health effects* (as opposed to thermal comfort) of extreme weather. These include the development of indices such as the heat index and the wind chill index that attempted to quantify the severity of heat and cold, respectively, in the outdoor environment. Works such as those by Gold (1935) and Siple and Passel (1975) helped form these indices. An outdoor health index

Table 3.4 Thermal comfort effect of land use categories

Land-use category	*Thermal comfort conditions (in TCI)*			
	Morning	*Early afternoon*	*Late afternoon*	*Night*
Low-density residential	4.8	8.0	1.3	0.6
Medium-density residential	1.8	4.8	3.4	1.3
Shopping area	3.6	6.8	5.8	−4.6
Central Business District	2.5	7.1	2.2	−4.7
Office complex	1.0	3.4	2.6	0.8
Heavy industrial	3.6	6.6	−1.9	−1.1
Park	1.0	3.4	2.1	0.8
Open grassland	5.8	6.3	3.6	−5.7
Open asphalt	5.8	11.7	4.9	−5.7

Source: Morgan and Baskett (1974)

Note: Thermal comfort is measured in "thermal comfort index" (TCI): +10 = unpleasantly hot; 5 = pleasantly warm; 0 = neutral; −5 = pleasantly cool; −10 = unpleasantly cold

widely used in the USA, New Zealand and Australia is the apparent temperature (AT) index developed by Steadman (1979, 1984). AT indicates the temperature which a given combination of dry-bulb temperature and vapor pressure is "supposed to feel like."

However, urban thermal comfort (as opposed to outdoor climate health) began receiving research interest only in the 1970s. These attempts used a man–environment energy balance approach and developed indices for thermal sensation as well as for thermal comfort (usually called "pleasantness"). An important pioneer of this approach is the "Manmo" model developed by Myrup and Morgan (1972). The "Manmo" model is based on a comfort index called the "thermal comfort index" (TCI) for the outdoors. Table 3.4 shows an example of the application of "Manmo" to the urban outdoors in a mid-latitude city in the summer.

The energy balance approach to urban thermal comfort has proceeded much the same way as the indoor thermal comfort, with suitable modifications to account for direct solar radiation (see, for example, Penwarden, 1973; Jendritzky and Nübler, 1981). However, there are several reasons why the indoor model is not conceptually suitable for outdoor thermal comfort. Höppe (2002) identified three basic reasons for the difference:

- psychological;
- thermophysiological;
- heat balance differences.

The psychological reasons for the difference between indoor and outdoor comfort perceptions have to do with expectation. People tolerate a larger variation in climatic conditions in the outdoors than the indoors, provided the outdoors has possibilities for adaptive behavior and, more importantly, affords suitable sociable spaces. This has been seen in beach resorts (Höppe and Seidl, 1991), urban parks and street canyons (Nikolopoulou *et al.*, 2001) where people did not mind warmer-than-usual conditions, as long as sociable spaces are accessible to them. Thus, we see urban gatherings for sports, carnivals, etc., even during the hot part of the day in the tropics, where crowds tend to ignore the excessive thermal stress on account of the ambience created by these events.

Outdoor spaces present few constraints. People use them of their own free choice: to soak in the sun, to get some fresh air, or to see and to be seen. Unlike the indoors, environmental stimulation is crucial to outdoor comfort. People want to "charge up" with warmth and fresh air, especially when considered in combination with their immediate thermal history (where they come from). This enables them to tolerate larger variation in the outdoor climate than the indoor.

Thermophysiological differences between indoor and outdoor comfort stem from differences in clothing, activity levels and exposures times. In warm climates, people tend to wear less clothing, do lighter activities and are exposed to environmental conditions longer in the indoors than outdoors. Exposure to outdoor climate is usually in the range of minutes, while indoor exposures last hours.

Thermal adaptation of the body to warm conditions is much faster than for cold conditions. Since the human body is constantly attempting to lose heat, it is able to adapt to hot conditions much faster. Yet, even in warm conditions, the physiological processes associated with the outdoors are vastly different from those of the indoors, particularly if outdoor conditions are more complex (say, a person walking through a street canyon with a complex mix of shaded and sunny areas) (see Höppe, 2002: 663–664). Since the human body is relatively large, there is a lag effect which may lead to a mismatch between thermal comfort and the shade/sunny patterns in the outdoors, compounding the quantification of outdoor thermal comfort. For example, the thermal sensation of a person walking through a sunny urban area and then resting in a shaded urban pocket for a few minutes may not exactly correspond with his/her activities, clothing, or even the micro-environmental conditions on account of the lag effect (see Nickolopoulou *et al.*, 2001).

The third reason why outdoor thermal comfort is different from the indoors stems from the heat balance differences between the two. While steady state conditions are possible in the indoors, they are rarely feasible in urban situations.

We can therefore say that the primary reasons for the difference between indoor and outdoor comfort is the energetic differences between the two. But that alone does not explain all the differences. There are different *expectations* from the two environments. People use the outdoors based on their perception of environmental variables (air temperatures, shade/sunny patterns, wind patterns, etc.), but once they have decided to come outdoors their expectations are different from that of the indoors. A purely heat balance/thermophysiological model would be unable to account for the difference. A combined psycho/heat balance approach may be necessary to explain and control urban thermal comfort (see Nikolopoulou *et al.*, 2001).

Outdoor comfort indices

The preceding section indicated the basic reasons for the differences in perception between the indoor and the outdoor climate. It therefore goes without saying that quantification of outdoor thermal comfort will have to chart its own path.

The quantification of urban thermal comfort is relatively new (less than thirty years old). Most of the efforts so far have utilized an energy model of some sort (Nickolopoulou and Steemers, 2003).

Wet-bulb globe temperature (WBGT) index

The WBGT is a thermal stress index that emphasizes the physiological effects of direct radiation and air movement. It is therefore suitable for stressful outdoor environments (both cold and heat stress).

The ISO 7243 (1989) suggests the following model for the WBGT index:

$$WBGT = 0.7T_w + 0.2T_g + 0.1T_a \tag{10}$$

T_w = temperature of a *naturally ventilated* wet-bulb thermometer (i.e. not aspirated);
T_g = temperature of a 150 mm diameter matt-black globe thermometer (°C);
T_a = air temperature (°C).

Table 3.5 WBGT reference values

Metabolic rate - M(watts)	*WBGT values (°C)*			
	Acclimatized		*Not acclimatized*	
Seated (66 < M < 130)	30		29	
Light walking (131 < M < 200)	28		26	
Brisk walking (201 < M < 260)	25	(26)	22	(23)
Heavy work (M > 261)	23	(25)	18	(20)

Source: Modified after ISO 7243 (1989).

Note: See discussions in text for an explanation of the reference values. Figures in parentheses are for situations with sensible (>0.15 m/s) air movement; other figures are for no sensible air movement (<0.15 m/s)

Table 3.5 presents the ISO 7243 reference WBGT values for various metabolic rates. The reference value for a given activity level is the WBGT at which the rectal temperature of persons carrying out that activity reaches a maximum of 100°F.

Since the WBGT is a stress index, it has been adopted by the National Institute of Occupational Safety and Health (NIOSH, 1986), ISO 7243 (1989), and by the US Navy for determining suitable times for outdoor training. Its major shortcomings are twofold. WBGT is not a comfort index. Furthermore, the data requirements are highly idiosyncratic (especially the T_g which has to be custom-measured) and therefore not readily available.

Physiological Equivalent Temperature (PET)

PET is defined as the physiologically equivalent temperature at any given place (outdoors or indoors) equivalent to the air temperature at which the actual skin temperature in a real environment is equal to that of a reference human body (Mayer and Höppe, 1987). The reference human body is defined as a person doing light activity (80 W above the basic metabolism) and wearing a clothing level of 0.9 clo.

The major shortcoming of the PET, especially from a tropical perspective, is its limits of applicability (air velocity <0.1 m/s and vapor pressure <12 hPa).

Outdoor Standard Effective Temperature (OUT_SET)

This approach takes a widely used and well-validated indoor thermal comfort index (SET*) and adapts it to account for solar and infra-red radiation (Pickup and de Dear, 2000; de Dear and Pickup, 2000; Spagnolo and de Dear, 2003). SET* is defined as the temperature of hypothetical isothermal reference environment (air temperature = mean radiant temperature; relative humidity = 50 percent; air velocity = 0.15 m/s) such that a person in the reference environment wearing 0.6 clo and standing still (met = 1.2) has the same mean skin temperature and skin wettedness as the person in the actual complex environment. In the outdoor version (OUT_SET), the assumption that MRT = AT is relaxed, and the actual MRT is calculated using a human thermoregulatory model. Currently, the OUT_SET model is being experimentally validated.

The derivation of PET and OUT_SET heavily depends on the correct estimation of the mean radiant temperature (T_{MRT}). Recently, Matzarakis *et al.* (2000) have developed computer simulation software for this purpose (RayMan) that takes into account short- and long-wave radiation fluxes as well as cloud-cover effects.

3.4 Bio-climatic implications of urban climate changes

While research on urban thermal comfort is scarce, research on comfort alterations due to the heat-island effect is practically non-existent. However, some research exists on thermal comfort in the outdoors in temperate cities, and this may provide a useful hypothesis about urban thermal comfort in the urban tropical context (Ahmed, 2003; Givoni *et al.*, 2003).

Thermal comfort in the temperate outdoors is due to two types of urban climate influences: thermal and mechanical – the latter being the comfort influence of wind flow. Influence of mechanical action of wind is greater upon cold discomfort than warm discomfort. Therefore, its role can be neglected in the tropics.

Based on this definition of urban thermal comfort, it can be seen that radiant temperature is critical to human comfort in the tropical outdoors (Lewis *et al.*, 1971; Plumley, 1977). Fanger alluded to this when he said that "the more non-uniform a thermal field [is], the greater the number of dissatisfied occupants" (Fanger, 1970: 218). This is borne by the following studies in the temperate zone.

Herrington and Vittum (1977) studied urban thermal comfort under cooler conditions (air temperatures less than 75°F or 23.9°C) at Syracuse, New York. They found little discomfort due to urban variations in air temperature or humidity. The presence or absence of wind, too, did not significantly influence thermal discomfort. However, variations in solar radiation input and long-wave radiation loss in the outdoors were significant factors in thermal comfort/discomfort in urban areas at all times of the day. The blocking off of sun in the urban canyon led to cool canyon surfaces which acted as heat sinks, drawing heat away from the body.

Stark and Miller (1977) conducted a similar study at Storrs, Connecticut, to ascertain summer-time thermal comfort in urban areas. This study, too, found little evidence for thermal discomfort due to urban variations in air temperature and wind speed. Significant differences in comfort, however, were found to be related to surface characteristics. Urban sites with substantial views to the sky and/or green surfaces had the lowest thermal stress. Locations with views predominantly to man-made surfaces experienced the highest thermal load. It was also shown that the percentage view of man-made surfaces was linearly related to net heat load on the human body ($R^2 = 0.87$).

These studies help us hypothesize that human discomfort in the tropics will be closely related to the radiation regime. This contrasts markedly with indoor thermal comfort where radiation is only a minor influence due to the negligible differences between indoor surface temperatures and air temperatures. Thus it is safe to assume that radiant temperature in the indoors is equal to the air temperature. In the urban environment, however, radiant temperatures vary according to shade, urban canyon geometry, vegetation availability and the seasons. Reduction in solar radiation availability in winter, and increase in the same during the summer, appears to cause the greatest variation in human thermal comfort in urban areas.

It is thus possible to postulate that the negative impact of urban climate warming in the tropics could be ameliorated somewhat by inducing evaporation, but a more pragmatic approach would be to reduce the potential for heating in the urban canopy layer (UCL). This will pre-empt the need for evaporative heat loss and will lead to lower ambient air temperature. Since heat loss at lower ambient air temperatures is easily and more pleasantly achieved by radiative and convective heat losses, such a strategy will also lead to greater urban thermal comfort. In this light, increase in natural surface cover in urban areas appears to be important.

However, other design strategies will also be necessary. In the outdoors it will be necessary to provide for some ventilation that could compensate the short environmental exposures people typically have. The question of rain protection, too, is of importance in the tropics. A horizontal wind flow of 7 m/sec may drive rain at 45°, while a fine drizzle may be so driven by winds as low as 1 m/sec (see, for example, Arens and Ballanti, 1977, Chand and Bhargava, 2002).

Considering the complexity of the issues involved, Givoni (1989: 7–5, 7–6) lists the following urban design objectives for improving outdoor comfort in the changing urban climate of tropical cities:

1. Provide shade for streets and outdoor activities.
2. Enable natural ventilation for urban spaces.
3. Minimize flood hazard.
4. Provide rain protection for pedestrians.
5. Regulate rain water flow.

The primary goal in urban design in the tropics is shading. Every other design decision must facilitate shading and augment the cooling so obtained by other means. Ventilation could play an important part in this regard. Night-time comfort must receive special attention in the equatorial tropics. Since nights in the urban tropics are typically too warm for comfort, (Nieuwolt, 1986: 532) heat storage in the urban built mass must be kept minimal.

In terms of urban design strategies, the approach to tropical urban comfort may be summarized as follows (Givoni, 1989: 7-6):

1. Street layout and system of network.
2. Density of built-up area.
3. Average and relative height of buildings.
4. Proper choice of building types (thick, thin, doughnut, etc.).
5. Design details of green spaces.

Bio-climatic influences of trees and buildings

Trees affect urban microclimates on two levels: human comfort and building energy budget (Miller, 1988). It is not only what vegetation does that is important but also what it prevents – the heating-up of urban canyons (Oke, 1989). While the direct temperature reduction due to the presence of a large urban park is only about 1–2°C (1.8–3.6°F), the immeasurable effect on reducing heat build-up, etc., will be of greater value (Oke, 1989; Shashua-Bar and Hoffman, 2003) (see also Table 2.5).

Part of the reason why vegetation is so effective in radiation control is the porosity of tree canopy (Heisler, 1974). A large amount of radiation impinging upon vegetation is retained to be used for transpiration. About 10–25 percent is used to heat the air. In contrast, open, paved urban areas like parking lots absorb up to 85 percent of the incident solar radiation and the rest is directly used to heat the air. This absorbed heat, too, will eventually heat the air. Myrup (1969) has estimated that urban areas with no vegetation would partition 40 percent of net all-wave radiation into sensible heat and the balance into heat storage, which in turn will add to sensible heat gain after sunset. Urban areas with significant tree population, on the other hand, partition only 10 percent of net all-wave radiation into sensible heat. Another 10 percent is stored in the bio-mass while the rest is partitioned into latent heat loss. Evaporation by vegetation is much higher since the leaves of a tree crown usually have five to seven times more area than the maximum horizontal cross section of the tree crown (Leonard, 1972).

Using the stored energy, vegetation can also transpire efficiently. Even an isolated tree can transpire up to 100 US gallons (379 liters) of water per day (Kramer and Kozlowski, 1960). This is equivalent to the cooling provided by five 20,000 BTU (5800 W/m^2) air conditioners running 20 hours per day (Federer, 1971). What trees in urban areas provide is relief from solar radiation by a combination of absorption and evaporation, not so much relief from air temperature per se.

How do the shading effects of building façades compare with those of vegetation? While vegetation provides two kinds of thermal relief (shading and therefore radiation reduction, as well as evaporative heat suppression), buildings are capable of shading alone. However, the thermal comfort effects of shading are independent of the source of shading, whether vegetational or building façade. This is borne out by Figure 3.8 where air temperatures in Singapore above a shaded portion of a street and under a shaded tree appear quite similar. On the other hand, the temperature above a sunny green lawn is as equally hot as that of an exposed asphalt street. Thus, from the perspective of urban thermal comfort, *shading is key* – and it does not matter where the shade comes from, vegetation or buildings.

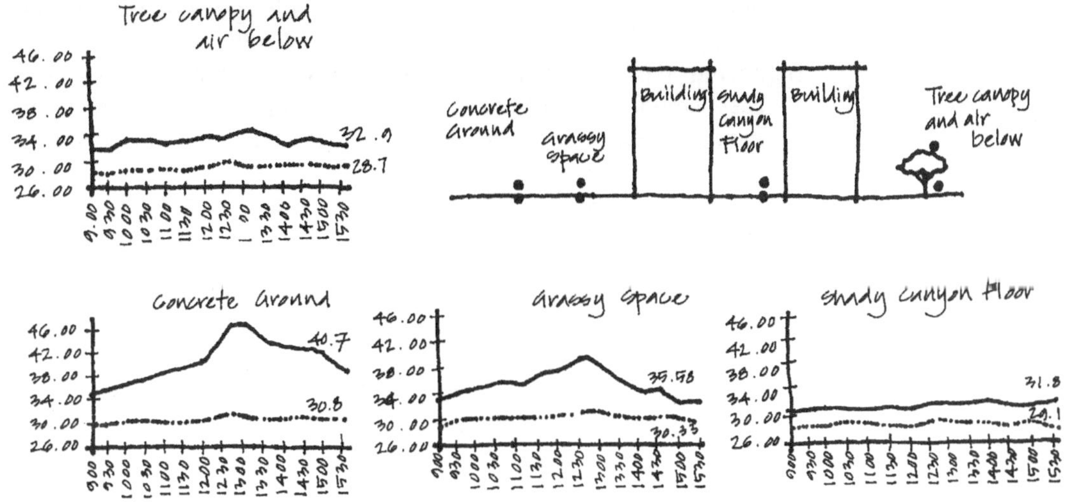

3.8 *Air temperature effects of shading: buildings and vegetation*

References

Ahmed, K.S. (2003). "Comfort in urban spaces: defining the boundaries of outdoor thermal comfort for the tropical urban environments," *Energy and Buildings*, 35(1): 103–110.

Ambler, H.R. (1955). "Notes on the climate of Nigeria with reference to personnel," *Journal of Tropical Medicine and Hygiene*, 58: 99–112.

Arens, E. and D. Ballanti (1977). "Outdoor comfort of pedestrians in cities," *Proceedings of a Conference on the Metropolitan Physical Environment*, Syracuse, N.Y., 25–29 August 1975, USDA Forest Service Technical Report NE-25, Upper Darby, Pa.: Northeastern Forest Experiment Station, pp. 115–129.

Arens, E., L. Zeren, R. Gonzales, L. Berglund, and P.E. McNall (1981). "A new bioclimatic chart for environmental design," in *Building Energy Management*, Proceedings of the ICBEM Conference, Pavoa de Varzim, 1980, London: Pergamon, pp. 645–657.

Arens, E.A. and N.S. Watanabe (1986). "A method for designing naturally cooled buildings using bio climate data," *ASHRAE Transactions*, PO-86-15, No. 1.

ASHRAE (1985). *Handbook of Fundamentals*, Atlanta, Ga: American Society of Heating, Refrigerating and Air-conditioning Engineers (ASHRAE).

ASHRAE (1992). *ANSI/ASHRAE 55-1992, Standard 55: Thermal Environmental Conditions for Human Occupancy*, Atlanta, Ga.: ASHRAE.

ASHVE (1932). *ASHVE Guide*, Atlanta, Ga: American Society of Heating and Ventilating Engineers (ASHVE).

Auliciems, A. (1983). "Psycho physical criteria for global thermal zones of building design," *Biometeorology*, 8(2): 69–86.

Ballantyne, E.R., J.R. Barned, and J.W. Spencer (1967). "Environment assessment of acclimatized Caucasian subjects at Port Moresby, Papua," *Proceedings of the Third Australian Building Research Congress*.

Basnayake, V. (1984). "Seasonal and diurnal variation in thermal comfort in Sri Lanka," *Journal of the National Science Council of Sri Lanka*, 12(2): 191–203.

Busch, J.F. (1992). "A tale of two populations: thermal comfort in air conditioned and naturally ventilated offices in Thailand," *Energy and Buildings*, 18: 235–249.

Chand, I. and P.K. Bhargava (2002). "Estimation of driving rain index for India," *Building and Environment*, 37: 549–554.

Correa, C. (1989). *The New Landscape: Urbanization in the Third World*, London: Butterworth Architecture.

De Dear, R.J. (1998). "A global database of thermal comfort field experiments," *ASHRAE Transactions*, 104(1): 1141–1152.

De dear, R.J. and G.S. Brager (1998). "Developing an adaptive model of thermal comfort and preference: field studies of thermal comfort and adaptation," *ASHRAE Transactions*, 104(1): 145–167.

De Dear, R. and G.S. Brager (2001). "The adaptive model of thermal comfort and energy conservation in the built environment," *International Journal of Biometerorology*, 45: 100–108.

De Dear, R.J. and G.S. Brager (2002). "Thermal comfort in naturally ventilated buildings: revisions to ASHRAE Standard 55," *Energy and Buildings*, 34(6): 549–561.

De Dear, R.J. and M.E. Fountain (1994). "Field experiments on occupant comfort and office thermal environments in a hot-humid climate," *ASHRAE Transactions*, 100(2): 457–475.

De Dear, R.J. and J. Pickup (2000). "An outdoor thermal comfort index (OUT_SET*) – Part II: Applications," in R.J. de Dear, J.D. Kalma, T.R. Oke, and A. Auliciems (eds), *Biometeorology and Urban Climatology at the Turn of the Millennium*, Selected Papers from the ICB-ICUC'99 Conference, Sydney, WCASP-50, WMO/TD No. 1026, pp. 285–290.

De Dear, R.J., G.S. Brager, and D. Cooper (1998). *Developing an Adaptive Model of Thermal Comfort and Preferences*, Final Report on ASHRAE RP-884, Atlanta, Ga.: ASHRAE.

De Dear, R.J., K.G. Leow, and A. Ameen (1991a). "Thermal comfort in the humid tropics – Part I: Climate chamber experiments on temperature preferences in Singapore," *ASHRAE Transactions*, 97(1): 874–879.

De Dear, R.J., K.G. Leow, and S.C. Foo (1991b). "Thermal comfort in the humid tropics: field experiments in air conditioned and naturally ventilated buildings in Singapore," *International Journal of Biometeorology*, 34: 259–265.

de Figueroa, R. (1980). *Architecture for the Tropics: A Bibliographical Synthesis (from the beginnings to 1972)*, Rio Piedras, Puerto Rico: Editorial Universitaria, University of Puerto Rico.

Drysdale, J.W. (1952). *Designing Houses for Australian Climates*, Bulletin No. 6, Experimental Building Station, Australian Government Publication Service.

Ellis, F.P. (1952). "Thermal comfort in warm, humid atmospheres. Observations in a warship in the Tropics," *Journal of Hygiene*, 50: 415.

Ellis, F.P. (1953). "Thermal comfort in warm and humid atmospheres. Observations on groups and individuals in Singapore," *Journal of Hygiene*, 51: 386.

Fanger, P.O. (1970). *Thermal Comfort: Analysis and Applications in Environmental Engineering*, New York: McGraw-Hill.

Fanger, P.O. and J. Toftum (2002). "Extension of the PMV model to non air-conditioned buildings in warm climates," *Energy and Buildings*, 34(6): 533–536.

Federer, C.A. (1971). "Effects of trees in modifying urban microclimate," in S. Little and J.H. Noyes (eds), *Trees and Forests in an Urbanizing Environment*, Amherst: University of Massachusetts Coop. Extension Service, pp. 23–28.

Gagge, A.P., A.P. Fobelets, and L.G. Berglund (1986). "A standard predictive index of human response to the thermal environment," *ASHRAE Transactions*, 92(2B): 709–718.

Givoni, B. (1989). *Urban Design in Different Climates*, World Meteorological Organization Technical Note No. 346, Geneva: World Meteorological Organization.

Givoni, B., M. Noguchi, H. Saaroni, O. Pochter, Y. Yaacov, N. Feller, and S. Becker (2003). "Outdoor comfort research issues," *Energy and Buildings*, 35(1): 77–86.

Gold, E. (1935). "The effect of wind temperature, humidity and sunshine on the loss of heat of a body at temperature 98°F," *Quarterly Journal of the Royal Meteorological Society*, 63(319).

Heisler, G.B. (1974). "Trees and human comfort in urban areas," *Journal of Forestry*, 72: 466–469.

Herrington L.P. and J.S. Vittum (1977). "Human thermal comfort in urban outdoor spaces," *Proceedings of a Conference on the Metropolitan Physical Environment*, Syracuse, N.Y., 25–29 August 1975. USDA Forest Service Technical Report NE-25, Upper Darby, Pa.: Northeastern Forest Experiment Station, pp. 130–138.

Heschong, L. (1979) *Thermal Delight in Architecture*, Cambridge, Mass.: MIT Press.

Höppe, P. (2002). "Different aspects of assessing indoor and outdoor thermal comfort," *Energy and Buildings*, 34(6): 661–665.

Höppe, P. and H.A.J. Seidl (1991). "Problems in the assessment of the bioclimate for vacationists at the seaside," *International Journal of Biometeorology*, 35: 107–110.

Humphreys, M.A. (1978). "Outdoor temperatures and comfort indoors," *Building Research and Practice* (Journal of the CIB), 6(2): 92–105.

Humphreys, M.A. (1981). "The dependence of comfortable temperatures on indoor and outdoor climates," in K. Cena and J.A. Clark (eds), *Bioengineering, Thermal Physiology and Comfort*, Amsterdam: Elsevier, pp. 229–250.

Humphreys, M.A. and J.F. Nicol (1995). "An adaptive guideline for UK office temperatures," in M.A. Humphreys, J.F. Nicol, O. Sykes, and S. Roaf (eds), *Standards for Thermal Comfort, Indoor Air Temperature Standards for the 21st Century*, London: E. & F.N. Spon, Chapter 20.

Humphreys, M.A. and J.F. Nicol (1996). "Conflicting criteria for thermal sensation within the Fanger Predicted Mean Vote Equation," *Proceedings of the CIBSE/ASHRAE Joint National Conference*, Harrogate.

ISO 7243 (1989). *Hot Environments – Estimation of the Heat Stress on Working Man, Based on the WBGT-Index (Wet Bulb Globe Temperature)*, Geneva, Switzerland: International Organization for Standardization.

ISO 7730 (1994). *Moderate Thermal Environments – Determination of the PMV and PPD Indices and Specification of the Conditions for Thermal Comfort*, (2nd edn), 1994-12-15, Reference Number ISO 7730:1994(E).

Jendritzky, G. and W. Nübler (1981). "A model analyzing the urban thermal environment in physiologically significant terms," *Archives for Meteorology, Geophysics and Bioclimatology, Series B*, 29: 313–326.

Jitkhajornwanich, K. and A.C. Pitts (2002). "Interpretation of thermal responses of four subject groups in transitional spaces of buildings in Bangkok," *Building and Environment*, 37: 1193–1204.

Khedari, J., N. Yamtraipat, N. Pratintong, and J. Hirunlabh (2000). "Thailand ventilation comfort chart," *Energy and Buildings*, 32: 245–249.

Kramer, P.J. and T.T. Kozlowski (1960). *Physiology of Trees*, New York: McGraw-Hill.

Kwok, A. (2000). "Thermal boredom," in the *Proceedings of the Passive and Low Energy Architecture Conference, PLEA 2000*, Cambridge, UK.

Leonard, R.E. (1972). "Making our lives more pleasant – plants as climate changers," in *Landscape for Living*, Washington, D.C.: USDA, pp. 5–9.

Lewis, J.E. *et al.* (1971). "Some effects of urban morphology on street level temperatures at Washington, D.C.," *Journal of the Washington Academy of Science*, 61: 258–265.

McIntyre, D.A. (1982). "Chamber studies – reductio ad absurdum?," *Energy and Buildings*, 5: 89–96.

Mather, J.R. (1974). *Climatology: Fundamentals and Applications*, New York: McGraw-Hill.

Matzarakis, A., F. Rutz, and H. Mayer (2000). "Estimation and calculation of the mean radiant temperature within urban structures," in R.J. de Dear, J.D. Kalma, T.R. Oke, and A. Auliciems (eds), *Biometeorology and Urban Climatology at the turn of the Millennium*, Selected Papers from the ICB-ICUC'99 Conference, Sydney, WCASP-50, WMO/TD No. 1026, pp. 273–278.

Mayer, H. and P. Höppe (1987). "Thermal comfort of man in different urban environments," *Theoretical and Applied Climatology*, 38: 43–49.

Miller, R.W. (1988). *Urban Forestry: Planning and Managing Urban Green Spaces*, Englewood Cliffs, N.J.: Prentice-Hall.

Monteith, J.L. (1973). *Principles of Environmental Physics*, London: Edward Arnold.

Morgan, D.L. and R.L. Baskett (1974). "Comfort of man in the city: an energy balance model of man-environment coupling," *International Journal of Biometeorology*, 18(3): 184–198.

Myrup, L.O. (1969). "A numerical model of the urban heat island," *Journal of Applied Meteorology*, 32: 896–907.

Myrup, L.O. and D.L Morgan (1972). *Numerical Model of the Urban Atmosphere*. Vol 1: *The City–Atmosphere Interface*, Contributions in Atmospheric Science No. 4, Davis, Calif.: University of California Press.

Nicol, J.F. (1974). "An analysis of some observations of thermal comfort in Roorkee, India, and Baghdad, Iraq," *Annals of Human Biology*, 1(4): 411–426.

Nicol, F. (unpublished). "Why International Thermal Comfort Standards Don't Fit Tropical Buildings," Oxford Centre for Sustainable Development, Oxford Brookes University, Oxford, UK.

Nicol, J.F. and M.A. Humphreys (2002). "Adaptive thermal comfort and sustainable thermal comfort standards for buildings," *Energy and Buildings*, 34(6): 563–572.

Nieuwolt, S. (1986). "Design for climate in hot, humid cities," in *Urban Climatology and its Applications with Special Regard to Tropical Areas*, Proceedings of a Technical Conference organized by the World Meteorological Organization, held in Mexico City, 26–30 November 1984, pp. 514–534.

Nikolopoulou, M., N. Baker, and K. Steemers (2001). "Thermal comfort in outdoor urban spaces: understanding the human parameter," *Solar Energy*, 70(3): 227–235.

Nikolopoulou, M. and K. Steemers (2003). "Thermal comfort and psychological adaptation as a guide for designing urban spaces," *Energy and Buildings*, 35(1): 95–101.

NIOSH (1986). *Occupational Exposure to Hot Environments*, National Institute for Occupational Safety and Health, DHHS (NIOSH) Publication No. 86–113, Washington, D.C.

Oke, T.R. (1987). *Boundary Layer Climates*, London: Methuen.

Oke, T.R. (1989). "The micrometeorology of the urban forest," *Philosophical Transactions of the Royal Society of London B*, 324: 335–350.

Olgyay, V. and A. Olgyay (1963). *Design with Climate: A Bio-climatic Approach to Architecture*, New York: McGraw-Hill.

Parker, J. (1983). "Do energy conserving landscapes work?," *Landscape Architecture*, 72(4): 89–90.

Penwarden, A. (1973). "Acceptable wind speeds in towns," *Building Science*, 8.

Pickup, J. and R.J. de Dear (2000). "An outdoor thermal comfort index (OUT_SET*) – Part I: The model and its assumptions," in R.J. de Dear, J.D. Kalma, T.R. Oke, and A. Auliciems (eds), *Biometeorology and Urban Climatology at the turn of the Millennium*, Selected Papers from the ICB-ICUC'99 Conference, Sydney, WCASP-50, WMO/TD No. 1026, pp. 279–283.

Plumley, H.J. (1977). "Design of outdoor spaces for thermal comfort," *Proceedings of a Conference on the Metropolitan Physical Environment*, Syracuse, N.Y., 25–29 August 1975, USDA Forest Service Technical Report NE-25, Upper Darby, Pa.: Northeastern Forest Experiment Station, pp. 152–162.

Prins, G. (1992). "On condis and coolth," *Energy and Buildings*, 18: 251–258.

Rao, M.N. (1952). "Comfort range in tropical Calcutta. A preliminary experiment," *Indian Journal of Medical Research*, 40: 45.

Rohles, F.H. (1978). "The empirical approach to thermal comfort," *ASHRAE Transactions*, 84(1): 725–732.

Rohles, F.H., J.E. Woods, and R.G. Nevins (1974). "The effects of air movement and temperature on the thermal sensations of sedentary man," *ASHRAE Transactions No. 2298*, 80(1): 101–119.

Roulet, C.-A. (2001). "Indoor environmental quality in buildings and its impact on outdoor environment," *Energy and Buildings*, 33: 183–191.

Shashua-Bar, L. and M.E. Hoffman (2003). "Geometry and orientation aspects in passive cooling of canyon streets with trees," *Energy and Buildings*, 35(1): 61–68.

Siple, P. and C. Passel (1975). "Measurements of dry atmospheric cooling in subfreezing temperatures," *Proceedings of the American Philosophical Society*, 89(177).

Spagnolo, J. and R. de Dear (2003). "A field study of thermal comfort in outdoor and semi-outdoor environments in subtropical Sydney Australia," *Building and Environment*, 38(5): 721–738.

Stark, T.F. and D.R. Miller (1977). "Effect of synthetic surfaces and vegetation in urban areas on human energy balance and comfort," *Proceedings of a Conference on the Metropolitan Physical Environment*, Syracuse, N.Y., 25–29 August 1975, USDA Forest Service Technical Report NE-25, Upper Darby, Pa.: Northeastern Forest Experiment Station, pp. 139–151.

Steadman, R. (1979). "The assessment of sultriness: Part I and II," *Journal of Applied Meteorology*, 18: 861–885.

Steadman, R. (1984). "A universal scale of apparent temperature," *Journal of Climate and Applied Meteorology*, 23: 1674–1687.

Szokolay, S.V. (2000). "Dilemmas of warm-humid climate house design: heavy vs. lightweight and cooling effect of air movement," in the Proceedings of the PLEA 2000 conference, *Architecture, City, Environment*, London: James and James, pp. 144–149.

Tanabe, S., K. Kimura, and T. Hara (1987). "Thermal comfort requirements during the summer season in Japan," *ASHRAE Transactions*, 93(1): 564–576.

Webb, C.G. (1952/53). "On some observations of indoor climate in Malaya," *Journal of the Institute of Heating and Ventilating Engineers*, 20: 189.

Webb, C.G. (1959). "An analysis of some observations of thermal comfort in an equatorial climate," *British Journal of Industrial Medicine*, 16: 297.

Wong, N.H., H. Feriadi, P.Y. Lim, K.W. Tham, C. Sekhar, and K.W. Cheong (2002). "Thermal comfort evaluation of naturally ventilated public housing in Singapore," *Building and Environment*, 37: 1262–1277.

Wyndham, C.H. (1963). "Thermal comfort in the hot, humid tropics of Australia," *British Journal of Industrial Medicine*, 20: 110.

Yaglou, C.P. and W.E. Miller (1924). "Effective temperature applied to industrial ventilation problems," *ASHVE Transactions*, 30: 339–364.

Climate-conscious urban design in the tropics CHAPTER 4

Basic principles

The discussions on urban thermal comfort in the previous chapter brought home the crucial differences between indoor and outdoor comforts. Essentially, these boil down to two features: excessive direct and diffuse radiation that make the basic assumption of indoor comfort models ($T_a = T_{MRT}$) untenable, and urban air movement patterns which are hard to predict yet significantly influence outdoor comfort. We further learnt that *shading* is the key to tropical urban thermal comfort; but this requirement ought to be augmented by the promotion of urban wind flow, however difficult it may be.

At the same time, the crucial *urban design* question remains to be answered: how can these facts be translated into design actions that are often piecemeal, and certainly carried out by different people acting at different places and times?

It may be appropriate to return to the context outlined in the introduction within which climate-sensitive urban designers in the tropics have to operate: high density, increasing urbanization, and altered microclimate. In the face of rapid urbanization and shrinking resources to cater to the urban needs, higher densities of urban growth are required of all tropical cities. That they should cool themselves by passive means in an ecologically sensible manner goes without saying. At the same time, the problem of housing dense and growing human conglomerations such as Djakarta, Manila, Accra, and Cartagena (Columbia) is not so much the lack of resources or space but one of rational allocation. If only the designers could calculate the actual intensity of current development patterns in these cities it would not be long before they discover that much denser growth is possible. It is also the most ecologically sound approach. Two new ways of looking at urban growth must however be accepted before such conclusions could be put to practical use.

The apparently high building density in the tropics is not so much due to high-rise buildings but because the occupancy rate per room is high (Correa, 1989: 42). Every available space is built upon illegally by urban migrants. There is a crying need for open space in tropical cities. Sub-divisioning of city land partitions even the little open space available into private, but thermally and ecologically insignificant, backyards. While open space in a temperate city like London is 3ha. per 1,000 people, Delhi has only 1.5ha. per 1,000 people. A tropical city like Bombay has only 0.1 ha. This paltry amount includes even the grass on traffic islands (Correa, 1989: 42).

Tay Kheng Soon (1993) argued that higher densities in the tropics are theoretically possible, provided *common amenities are adequately provided and access to common areas are ensured.* Herein lies an important conceptual direction for climate-conscious urban design in the tropics: the design of spaces between buildings in a climatically suitable way to ensure usability *and* accessibility.

In a landmark study conducted in temperate cities, March and Martin (1972) found that even the apparently dense North American cities have very low floor area ratios (FAR). They also showed that a city like New York could be developed to its present intensity with much smaller buildings (not more than six to eight floors) if the road networks could be reorganized. Such reorganization can be the second conceptual direction for a high density, low-rise, climate-conscious growth in the tropics: a rethinking of urban transportation.

Many European cities are in the process of rediscovering the possibilities of high-density, low-energy, more humane urban growth brought about by pedestrianization. Cities like Amsterdam, Stockholm, Athens (Havlick, 1983) and Copenhagen (Gehl, 1989) have reorganized city street networks so as to facilitate a climatically and aesthetically appropriate growth.

Even a motorized city like New York depends to a large part on pedestrianization. It is estimated that more than 70 percent of all rush-hour traffic in Manhattan is on foot (Wright, 1992). The tautology of mass transit and density ("mass transit will not work unless density is high: people will not live densely until a good mass transit system is established") can only be solved by moving in the direction of high-density development *and* pedestrianization simultaneously.

Once these two preconditions (design of accessible public spaces and pedestrianization) are met, a new approach to climate-conscious design in the urban tropics can be attempted. In other words, the spaces between buildings must be designed sensitive to climate, and the use of such spaces must be made part and parcel of daily urban living by ensuring accessibility. We call it the design of the "commons."

4.1 Design of the "commons"

The idea of the "commons" was first developed as a way of understanding human population dynamics by William Forster Lloyd in 1833

(Hardin and Badden, 1977). It was later developed as a notion of that which is opposed to the "private." In our age, "commons" refers to "that part of the environment that lay beyond a person's own threshold and outside his own possession, but to which, however, that person had a recognized claim of usage – not to produce commodities but to provide for the subsistence of kin" (Illich, 1982: 16). Illich's definition includes the sidewalks and semi-public areas of street where one can relate to other human beings. "Commons" is taken to mean all realities that are naturally endowed for the common good of all. In this sense, it would include physical attributes like meadows, scenic views, beaches, etc., and psycho-social attributes like common language, culture, caring for the elderly, sick, etc. In this chapter we will limit the notion of the commons to its spatial attributes only. Illich clarifies this understanding better:

> Just as the home reflects in its shape the rhythm and extent of family life, so the commons are the traces of the commonality. There can be no dwelling without its commons . . . Spaces fit to bear the marks of life [are] as basic for survival as clean water and fresh air. Human beings simply do not fit into garages, no matter how splendidly furnished with . . . energy saving devices (Illich, 1982: 60).

One of the problems with past attempts at climate-conscious design was the insensitivity to the qualitative attributes of design, especially at the urban scale. Energy-saving techniques are expected to be widely used simply because of their economic potential. Experience has shown that such assumptions do not always hold. The treatment of climate-sensitive design as an issue concerned with the commons will help humanize the commons. Furthermore, it is only at such scales (between buildings and neighborhoods) that climate-conscious design can effectively tackle the problems caused by urban heat islands while ensuring high-density living.

However, there are certain urban activities that best suit the commons and others that work only in the private. Correa (1989) identifies a hierarchy of four urban space usage patterns that are typical of the developing world: space needed by the family for private uses (cooking, sleeping, storage, etc.), intimate contact space with the outside world (the front doorstep), neighborhood meeting places (wells, city water tap), urban gathering places (*maidan*). The last three fall within the domain of the commons to varying degrees. Climate-conscious urban design initiatives in the urban tropics must concern themselves with these three kinds of urban space needs.

This is exactly what we seek to achieve in the present chapter: the design fulfillment of the above-mentioned urban space needs in a climatically sensitive manner. The emphasis is on shading the commons so that a greater flexibility of space usage is facilitated in tropical cities. The primary design strategy employed here is to identify those activities that are manifestations of the public space needs of urban dwellers, and to facilitate them in terms of interconnected activity patterns, for urban activities are not confined to single-function zones – a model so perverse in current urban growth planning.

The focus of the book is to develop patterns of design that facilitate climatically suitable urban form. We will also have to return to the question of climate-sensitive urban transportation. However, before these are attempted, we will look at two design principles that promote shading and urban wind flow in outdoor situations and outline the techniques of their quantification.

Tropical living is a part indoors and part outdoors activity, though one suspects mostly the latter. Living in the tropical outdoors is relatively pleasant for most of the year. Correa quantifies this aspect of flexibility of outdoor use by a factor called the "usability co-efficient" (Correa, 1989). While people spend longer at any given time indoors, outdoor occupants are typically there for shorter periods at a given time. As such, shading to reduce the thermal stress *and* perceivable air movement to enhance the cooling effect are necessary in the outdoors.

A sole reliance on wind movement alone is not tenable in the urban tropics. Although such a strategy is routinely touted as a panacea for all climatic ills in the tropics, we must remember two facts about tropical wind patterns. On the one hand, the night-time rate of air movement is low or nil in the tropics. Furthermore, being in the "doldrums" twice a year for extensive periods, the equatorial tropics do not have high wind velocities at the macro-level. Most daily wind flows are local in nature (like the sea–land breeze). A nocturnal urban heat-island tends to keep the air temperature over land almost as high as that over the sea, thus weakening any night-time land breeze.

The "problem" of climate-conscious tropical urban design is therefore twofold: prevention of heat build-up as the day unfolds (so as to reduce night-time cooling loads) and encouraging convective cooling at night.

Viewed in this manner, the urban design goals would be (a) radiation reduction during the day, and (b) ventilative cooling at night. The nature of the climatic problem is such that neither of these goals can be considered in isolation.

4.2 Solar radiation prevention: the "shadow umbrella"

While sun prevention at the building scale is fairly well known, what has largely been neglected is sun prevention at the urban scale. More precisely, the art of using urban *morphology* to create shaded public places has received little attention in the tropics.

In order to fill this gap we develop in this chapter a hypothetical "shadow umbrella" for radiation reduction in the *outdoors* during the day. The concept was first introduced by Emmanuel (1993). The primary tool in achieving shading is urban massing, although the potential of vegetation in enhancing shading and reducing the urban heat island is also considered. The use of water in achieving/enhancing night-time ventilative cooling is a secondary consideration. In doing so, the "shadow umbrella" concept addresses two design issues: the creation of shaded urban public places that have direct bearing on outdoor thermal comfort (see, for example, Nikolopoulou *et al.*, 2001; Emmanuel and Perera, 2002), and determining the location(s) for bodies of water (lakes, pools or fountains) within this overall "shadow umbrella." It is a concept exactly opposite to what Knowles (1981) called a "solar envelope." This envelope determines not only the buildable volume but also the area surrounding it – the "commons."

Development of a "shadow umbrella"

Although a detailed discussion of solar geometry is beyond the scope of this book, let us look at some basic concepts as they relate to building shading in order to develop an *urban* massing that can shade itself. For our purposes, we could say that solar shading depends on the following:

- date of year;
- time of day;
- location;
- building/site orientation and dimensions.

The fact that the earth's plane of rotation around the sun is different from that of its rotation around itself causes the earth to appear as if it is declined to/from the sun. This appearance is defined by "declination angle" which varies between –23.45 and +23.45°.

$$\delta = 23.45 \times \sin\left(\frac{360 \times [284 + D]}{365.25}\right) \qquad (1)$$

where δ is the declination angle and D is the Julian date of the day of the year for which the declination angle is being calculated (Julian date being the number of days between January 1 and a given date).

The second factor that influences solar geometry calculations is the time of day. Two corrections are necessary here. On the one hand, the speed at which the earth travels around the sun is not constant. On the other, the time of day indicated by our clocks is not exactly the solar time; the local time is an artificial construct, determined by political authority and largely based on custom and convenience. These two factors (variations in time as the earth orbits the sun, and differences between solar time and local time) are related thus:

$$AST = LST - ET - 4\,(LON - LSM) \qquad (2)$$

where *AST* = apparent solar time (i.e. solar time);
LST = local standard time (i.e. time indicated by our clocks);
ET = equation of time (get values from Table 4.1);
LON = longitude of a given site;
LSM = local standard meridian (i.e. difference between LST and Greenwich Mean Time [GMT] multiplied by 15°).

Having thus determined the declination angle of a given day and converted local time to solar time, we can now define the sun's position in the sky by two angles: solar altitude (β) and solar azimuth (ϕ). Solar altitude refers to the vertical angle between the horizon and the line of sight to the sun. Solar azimuth is the horizontal angle between the projection on the ground of the line of sight to the sun and the north–south axis (Figure 4.1). According to the conventions of the International Solar Energy Society (and the navigational charts), solar azimuth is expressed as the deviation in degrees from the *north* in a clockwise direction. In the US however, azimuth angles are expressed as deviations from the *south*:

$$\beta = \sin^{-1}(\cos L \times \cos \delta \times \cos H - \sin L \times \sin \delta) \qquad (3)$$

$$\phi = \cos^{-1}\left[\frac{\sin \beta \times \sin L - \sin \delta}{\cos \beta \times \cos L}\right] \qquad (4)$$

Table 4.1 Values for "equation of time" (in minutes and seconds)

	Day of month							
	1	*5*	*9*	*13*	*17*	*21*	*25*	*29*
January	3' 14"	5' 06"	6' 50"	8' 27"	9' 54"	11' 10"	12' 14"	13' 05"
February	13' 34"	14' 02"	14' 17"	14' 20"	14' 10"	13' 50"	13' 19"	
March	12' 38"	11' 48"	10' 51"	9' 49"	8' 42"	7' 32"	6' 20"	5' 07"
April	4' 12"	3' 01"	1' 52"	0' 47"	−0' 13"	−1' 06"	−1' 53"	−2' 33"
May	−2' 50"	−3' 17"	−3' 35"	−3' 44"	−3' 44"	−3' 34"	−3' 16"	−2' 51"
June	−2' 27"	−1' 49"	−1' 06"	−0' 18"	0' 33"	1' 25"	2' 17"	3' 07"
July	3' 31"	4' 16"	4' 56"	5' 30"	5' 57"	6' 15"	6' 24"	6' 23"
August	6' 17"	5' 59"	5' 33"	4' 57"	4' 12"	3' 19"	2' 18"	1' 10"
September	0' 15"	−1' 02"	−2' 22"	−3' 45"	−5' 10"	−6' 35"	−8' 00"	−9' 22"
October	−10' 01"	−11' 17"	−12' 27"	−13' 30"	−14' 25"	−15' 10"	−15' 46"	−16' 10"
November	−16' 21"	−16' 23"	−16' 12"	−15' 47"	−15' 10"	−14' 18"	−13' 15"	−11' 59"
December	−11' 16"	−9' 43"	−8' 01"	−6' 12"	−4' 17"	−2' 19"	−0' 20"	1' 39"

Source: Smithsonian Met. Tables, quoted by Watson & Labs (1983).

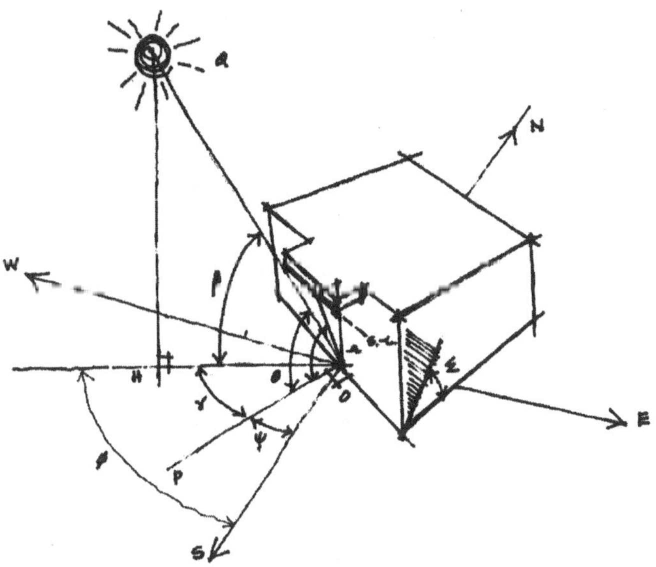

4.1 *Solar geometry angles*

	Angle	Between
β	Solar altitude	Solar rays ↔ horizontal plane
ϕ	Solar azimuth	Vertical projection of solar ray on horizontal plane ↔ south
θ	Incident angle	Solar ray ↔ normal to surface
ψ	Surface azimuth	Plane perpendicular to surface ↔ south
γ	Surface solar azimuth	Plane perpendicular to surface ↔ vertical Projection of solar ray on horizontal plane
Ω	Profile angle (shadow line angle)	Plane of solar rays passing through a horizontal shadow-line ↔ horizontal plane
Σ	Surface tilt angle	Tilted surface ↔ horizontal plane

where L is the local latitude and H is the hour angle. H is defined as the angle between the sun's current position and its position at solar noon. It is derived by multiplying the difference between current solar time and solar noon by 15°.

The sun's position in the sky casts shadows on the earth by a vertical or a horizontal element. A vertical element casts shadow due to a horizontal shadow angle while a horizontal element casts shadow using a vertical shadow angle (Figure 4.2). The use of either of these shadow angles in building shading depends on which angle necessitates the *smaller* shading device. The smaller shadow angle then becomes the selected angle for that particular location and that particular orientation.

Facades that face north or south can be easily shaded by horizontal

4.2 *Vertical and horizontal shadow angles*

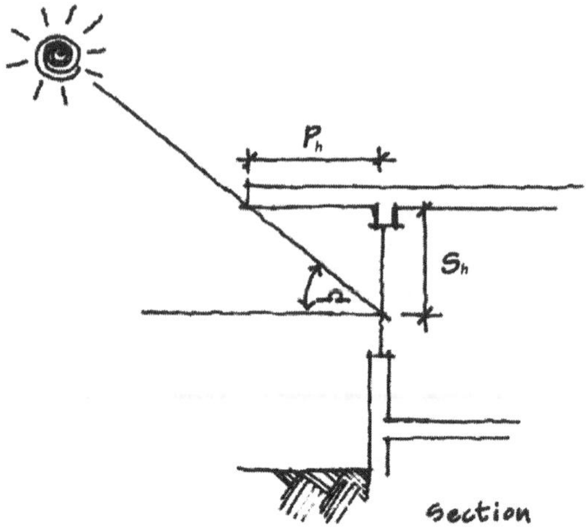

The depth of the horizontal projection P_h that results in a shadow height S_h can be calculated by

$$P_h = S_h * \cot \Omega$$

where,

Ω = Profile angle

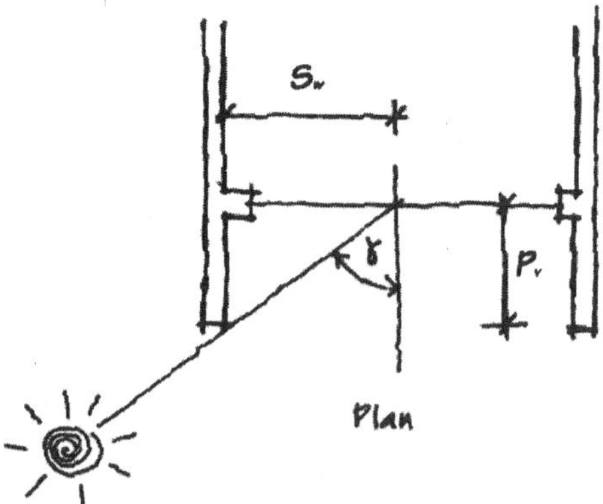

The shadow width cast Sw by a vertical fin with projection P_v can be calculated by

$$P_v = S_w * \cot \gamma$$

Or for the shadow width:

$$S_w = P_v * \tan \gamma$$

elements. The depth of a horizontal element (P_h) that casts a shadow height of S_h can be defined by:

$$P_h = \frac{S_h}{\tan \Omega} \tag{5}$$

Where Ω is the profile angle, which in turn depends on the solar altitude angle (β) and the difference between solar and surface azimuth angles (γ):

$$\tan \Omega = \frac{\tan \beta}{\cos \gamma} \tag{6}$$

Façades that face east or west cannot be effectively shaded by horizontal elements. In these instances, vertical elements are necessary. The depth of a vertical projection (P_v) that casts a shadow width of (S_w) can be defined by:

$$P_v = \frac{S_w}{\tan \gamma} \tag{7}$$

Using equations (1) to (7) it is possible to determine the devices necessary to shade any surface. Although these relate to shading of individual buildings, the principle is the same if one wants to extend shading to urban spaces. Just as shading elements attached to buildings can shade the openings in the building, urban spaces can be shaded by other buildings if they are favorably arranged.

The issue that matters most in terms of urban thermal comfort and urban geometry is the "view factor" (i.e. the proportion of the total spherical field of view from a subject taken up by surfaces). Another determinant is the temperature difference between the surfaces and the human being in the urban system: if surfaces are cooler, the human body loses heat to them; if warmer, the body receives heat. If urban surfaces could be kept "cool," people passing by would lose heat to them, and, in the tropical region, would therefore be comfortable. In keeping with this principle, the following steps are taken to establish an urban massing in a neighborhood that would keep its surfaces "cooler" (or shaded) in the tropics. The urban massing that can shade the immediate surroundings is what we call the "shadow umbrella."

The fundamental step in arriving at such a "shadow umbrella" is to establish the shadow angles. These angles vary by location (latitudes), time of day, and orientation. Since the purpose is to shade a particular surface at all times, the lowest angles must first be established.

Unlike in the temperate regions, the sun rays reach the tropical regions from all directions, including the north and the south. The northernmost solar exposure occurs during the summer solstice (June 21, or thereabouts). The southernmost exposure will be on the winter solstice (December 21, or thereabouts). While these two days determine the northern and southern extremities of sun positions, respectively, cut-off times on each of these days determine the eastern and western extremities.

The immediate problem is to establish the time limit within which shading is to be provided. In order to do this we will have to consider the general weather pattern in the tropics. The general air temperature patterns in the tropics indicate that temperature minima occur at around 06.00 hours and rapidly rise within the next three hours or so. By 09.00 hours, temperature reaches near maximum level and stays that way until a little before sundown. Thereafter it declines rapidly, starting at around 17.00 hours. Therefore, areas in the tropics where much activity occurs in the morning (say, commercial areas) should be kept cool – at least a little while *before* people congregate. In the case of residential zones, a later time limit could be set as internal migration of activities within houses is quite easily achieved. Shading until sunset, however, is needed in the tropics since the high thermal stress reached in early afternoon lingers on even after sunset.

We could therefore say that 08.00 hours will determine the eastern extremity and 17.00 hours. will determine the western extremity in commercial zones. The cut-off times for residential neighborhoods could be slightly relaxed (09.00–17.00 hours).

Figure 4.3 shows how these four positions of the sun (09.00 hours on June 21, 17.00 hours on June 21, 09.00 hours on December 21, and 17.00 hours on December 21) will determine the four corners of a perfectly shaded commercial city block. It turns out that the block will

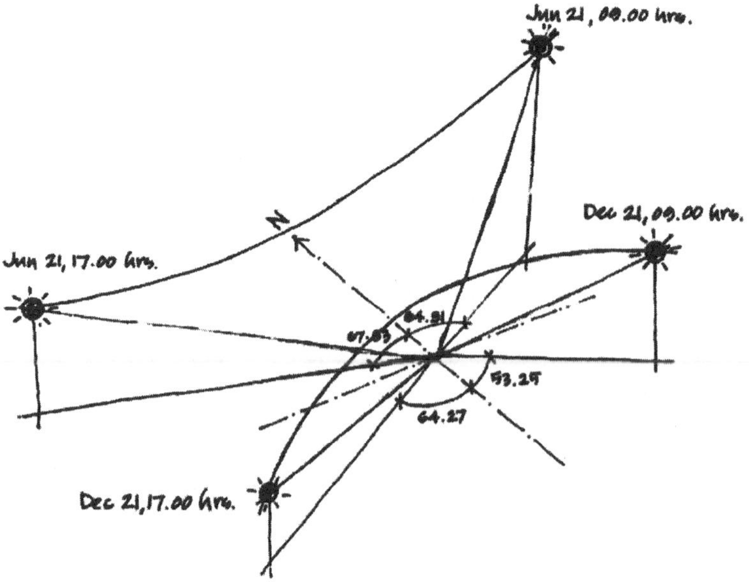

4.3 *Shadow cut-off times for the tropics*

Table 4.2 Selected shadow angles for a cut-off time of 08.00 hours. (solar time)

Orientation	*Latitude*					
	2	*4*	*6*	*8*	*10*	*12*
10	42	44	46	48	49	51
20	36	38	40	41	42	44
30	33	34	36	37	38	39
40	30	31	33	34	34	36
50	29	30	31	32	32	34
60	28	29	30	31	31	32
70	28	29	30	31	31	31
80	29	30	31	30	29	27
150	30	29	29	27	26	24
160	33	32	32	30	29	26
170	38	37	37	33	30	30
350	26	26	28	31	35	39
360	33	35	38	41	45	49

Notes:
For orientation the 350° and 360° angles shown are horizontal shadow angles. For all other orientations, angles refer to vertical shadow angles. Orientations not shown are deemed unsuitable for the tropics.

be elongated along north/south, and have a proportion of roughly 1:2 (east/west:north/south).

If buildings are to be erected on this block in such a way as to shade the entire block, the four positions of the sun can again be used to determine the building heights necessary for such shading. Figure 4.3 shows such a built massing. The sloping plain in the figure is the "shadow umbrella."

Tables 4.2 and 4.3 are the result of extensive use of computer software in determining shadow angles that would shade a given orientation at all times. The columns are latitudes (0°–12°, which is roughly the region of the tropics) and the rows are the orientations (from north) of the surfaces to be shaded. Orientations not shown are excluded since they demand too-low angles that would require unwieldy shading devices. The preferable orientations in the tropics are graphically represented in Figure 4.4.

With the shadow angles and preferred orientations thus established, the next step is to determine the plan dimension of a neighborhood block. Assuming that the east/west facing sides to be 30.48 m (100 ft), equations (1) to (7) yield the north/south facing side to be 75.44 m (247.5 ft) (Figure 4.5). The shadow angle used was that of the western orientation.

Assuming that an urban outdoor volume with a height equivalent to a two-storeyed building is to be shaded, the next step is to find out the

Table 4.3 Selected shadow angles for a cut-off time of 09.00 hours (solar time)

Orientation	*Latitude*					
	2	*4*	*6*	*8*	*10*	*12*
10	54	57	57	59	61	64
20	49	51	52	54	56	58
30	45	48	48	50	51	53
40	43	45	45	47	48	50
50	41	44	44	45	46	48
60	41	43	43	44	45	46
70	41	43	43	44	45	46
80	43	44	44	45	46	47
90	44	43	43	42	42	41
100	41	41	40	39	38	38
150	42	41	40	38	37	36
160	45	44	42	41	40	38
360	33	39	38	41	45	48

Notes:
* For orientation 360° angle shown is an horizontal shadow angle. For all other orientations, angles refer to vertical shadow angles. Orientations not shown are deemed unsuitable for the tropics.

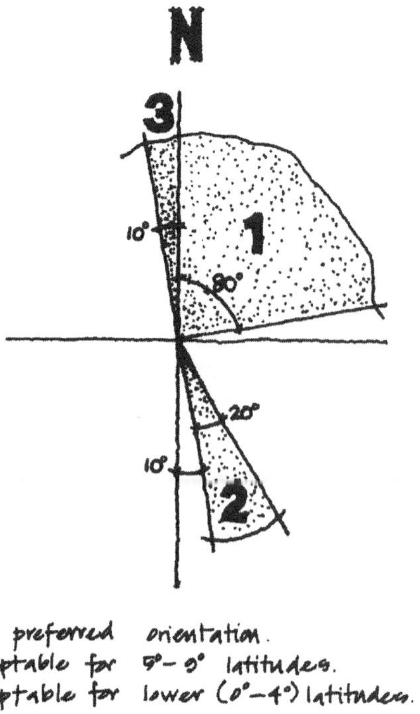

4.4 *Preferred orientations in the tropics*

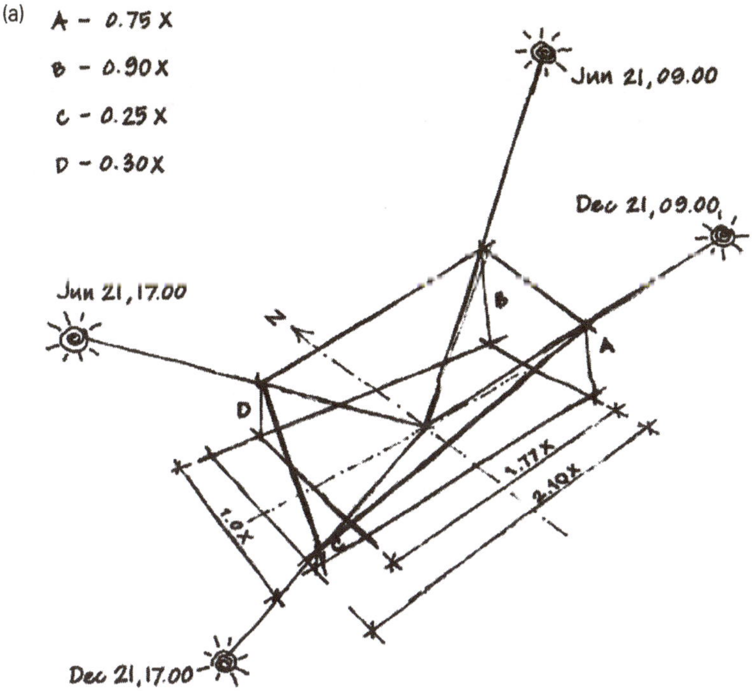

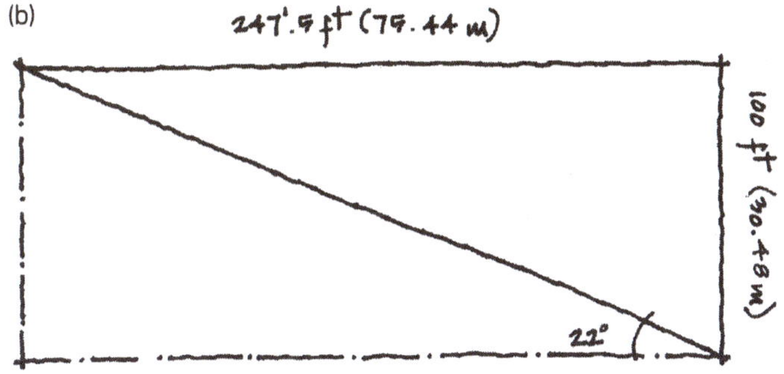

4.5 *An "ideal" shadeable urban block*

surrounding massing needed to shade the selected block in its entirety. Figure 4.6 shows the development of such a massing. The angles denoted are the shadow angles for orientations north (0°), east (90°), and so on.

Similarly, a "shadow umbrella" for a residential neighborhood could also be developed, this time using shadow angles that have a time limit of 09.00 hours to sunset (as opposed to 08.00 hours – sunset in the case of commercial areas) (see Figure 4.7).

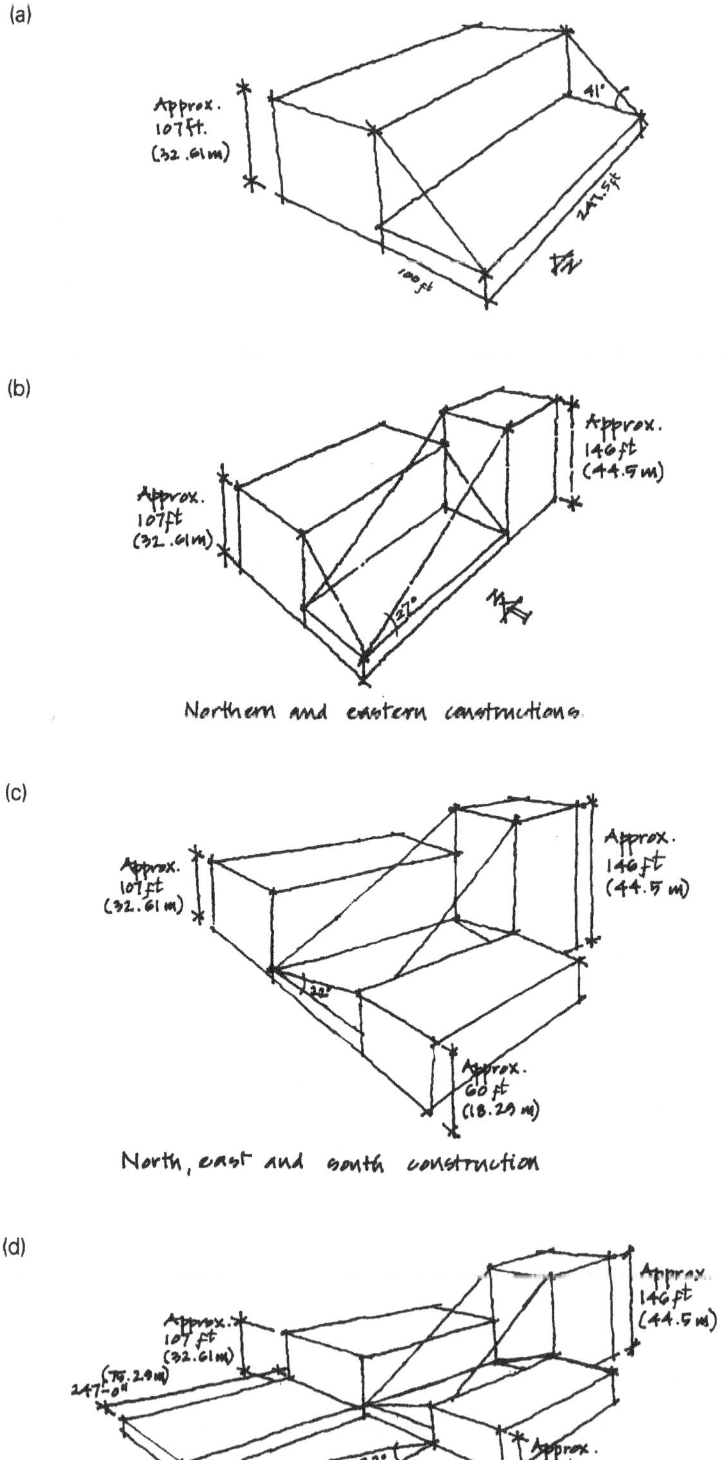

4.6 *Development of a "shadow umbrella": commercial area*

4.7 *Development of a "shadow umbrella": residential area*

(a)

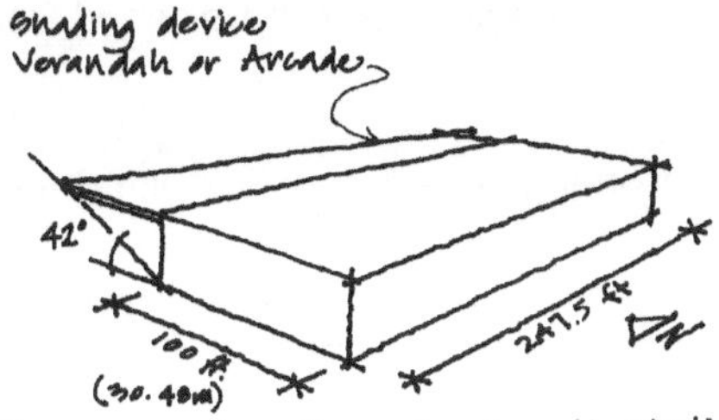

Northern construction, showing limitations to the northen edge of the buildable area.

(b)

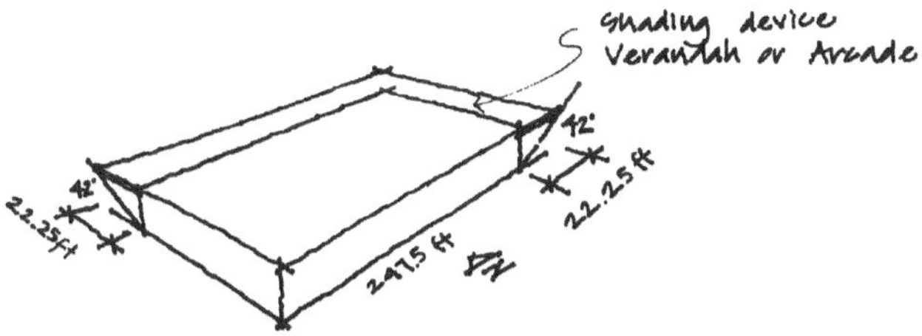

Northen and southern construction.
approximately 22 ft. (6.7m) "eaves" or protecting devices are necessary at the northern and eastern sides of the buildable area.

(c)

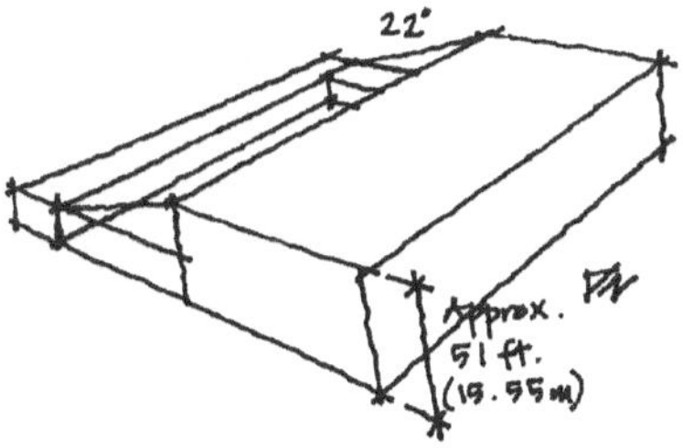

Northern, eastern and southern constructions.
A building (approximately five storeys high is needed to protect the proposed outdoor area

(d)

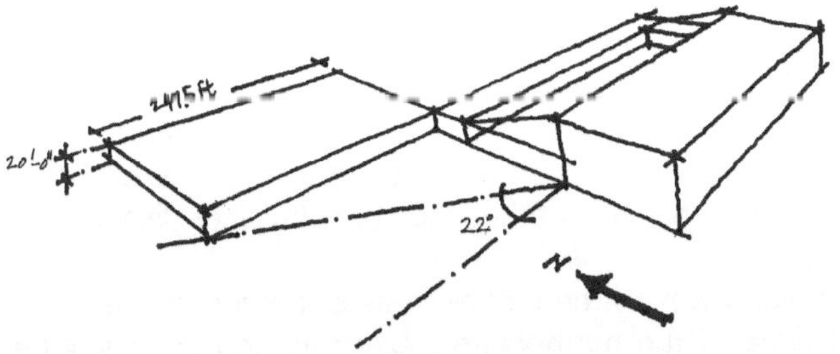

The following design strategies could be gleaned from the above examples:

1. The most preferred orientations are 0°–80° (i.e. within the north-eastern quadrant.
2. The urban density should be one that slopes towards the northeast (i.e. smaller buildings in the northeast and taller ones towards west).
3. Northeastern and southwestern corners of a development could be left open. Within this overall shadow massing, it is now possible to determine location/s of natural elements that would enhance the thermal comfort aspects of the outdoor areas in the whole neighborhood.

Use of natural elements as "tempering agents"

Figures 4.6(d) and 4.7(d) show the overall urban shadow configurations for commercial and residential blocks, respectively, in a tropical city. It is apparent that there are vast open spaces in both cases, offering ample opportunity to fine tune the "shadow umbrella."

The most peculiar climatic event in the tropics is the monsoon season, particularly in the equatorial tropics where two such monsoons differentiate the "seasons": northeast monsoon (December–January) and southwest monsoon (June–August). As can be seen from Figure 4.7(d), shadow massing developed so far shows compatibility with the general wind patterns in the area.

Carmona (1986) suggested that outdoor spaces in the tropics be so designed that the air stream passing through them to windward of each building is not unnecessarily overheated, thus nullifying the wind's beneficial effect. Noting that the equatorial winds are primarily northeasterly or southwesterly in origin and the shadow massing so far developed has open spaces at these sides, a case for locating waterbodies in the northeastern/southwestern quarters of the site can be made. Such a location would help cool the incoming air sufficiently so as to "temper" it and enhance its cooling potential (see Chapter 4 for discussion on air's cooling potential in the tropics).

Figure 4.8 shows the location of two ponds closer to the neighborhood (residential) block under consideration. Note also the verandah/arcade arrangements on the northern and eastern edges of the block. Thus a cool promenade encircling the lake is created which reinforces the traditional notion of a lake as the center in some of the tropical cities (for example, the historical hill capital of Sri Lanka – Kandy, and the ancient south-Indian temple cities like Madurai and Trichchirappalli).

Water alone will not be sufficient as a tempering agent, since the southern edge of the northwestern block in Figure 4.8 will be under

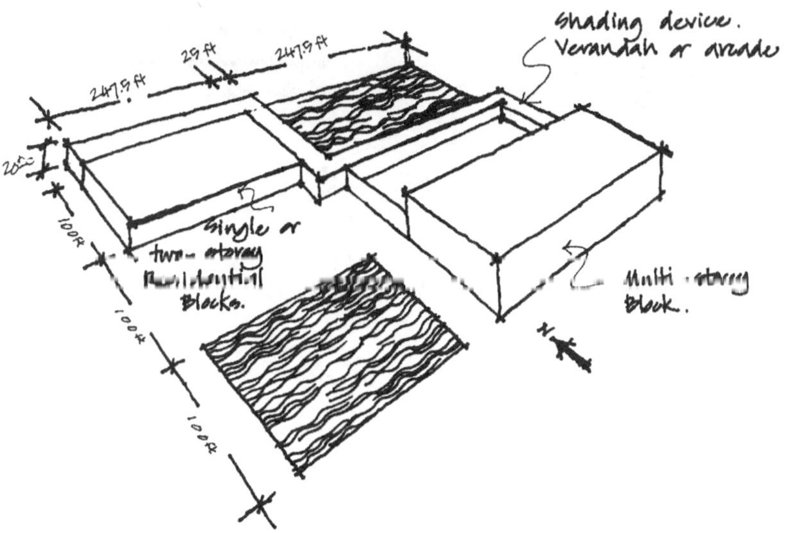

4.8 *A combined shadow envelope*

sun for some of the time. Trees could be planted as a buffer on this side.

One of the characteristics of the tropics is the weak temperature variation, both diurnal and seasonal. Larger climate-modifying elements like lakes are of little use to particular outdoor stretches if micro-level enhancements are not provided. In terms of thermal comfort in the tropics it is necessary to provide smaller comfort-enhancing elements in each urban block. Therefore, many smaller but climatically viable parks will ameliorate the microclimate over a wider area than one, large, centrally located park (Figure 4.9).

Working from the overall concept thus derived, a larger neighborhood could be designed in a climatically-appropriate way. Figure 4.10

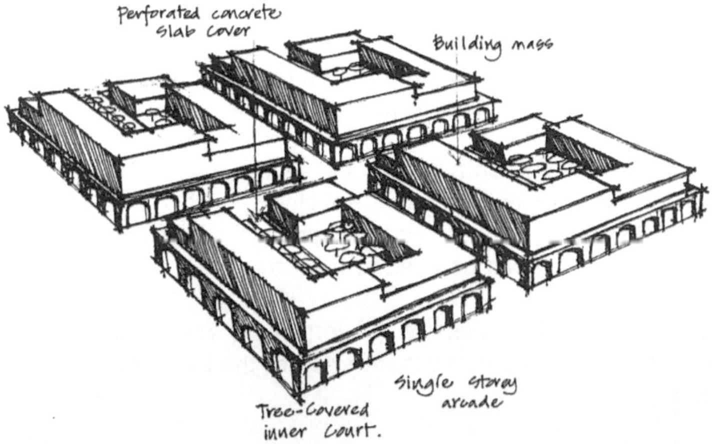

4.9 *"Shadow umbrella" for a commercial district*

4.10 *Composite "shadow umbrella"*

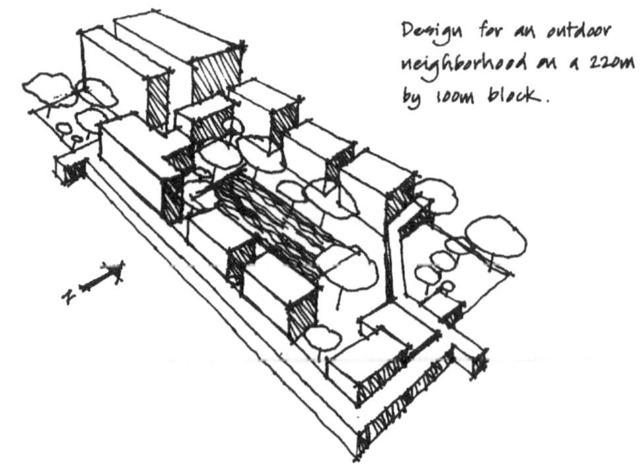

shows one such attempt on an approximately 220 m × 100 m area. It keeps all the principles thus far developed (smaller buildings on the northeast, taller buildings on the west, etc.) and leaves the center open for climate-ameliorating land uses. The pathways are so designed as to pick up the cool air from the centrally located pond and spread it along a northeast/southwest axis. Trees are used to reduce the heat radiation from large open areas where built mass is of lower height.

In summary, we can state that the "shadow umbrella" is a north-east to southwest sloping plane that is tall on the north and east and shorter on the south and west. The western side, which is the most difficult to shade in the tropical tropics (due to a later cut-off time), is almost empty and buildings on the side, if any, are very short (mostly a single-storey high). While the outer sides of buildings need to be shaded by shading devices, the inner sides of the buildings and the spaces between buildings are completely shaded by the built mass, within the cut-off times selected. Within this overall concept several alternative building massing options are possible (Figure 4.11). Notice the high floor area ratio (FAR) achieved in each of the alternative building arrangements proposed.

The following generic patterns can be detected from the "shadow umbrella" developed for the tropics:

1. Buildings will be tall on the north and east.
2. Buildings on the west will be the smallest.
3. It is preferable to leave a wide margin on the western side of the block. This space can be used to plant trees, etc., that could offer additional shading.
4. The center of the block can be left open. This opening can be either parallel to the north/south sides of the block, or a U-shaped court.

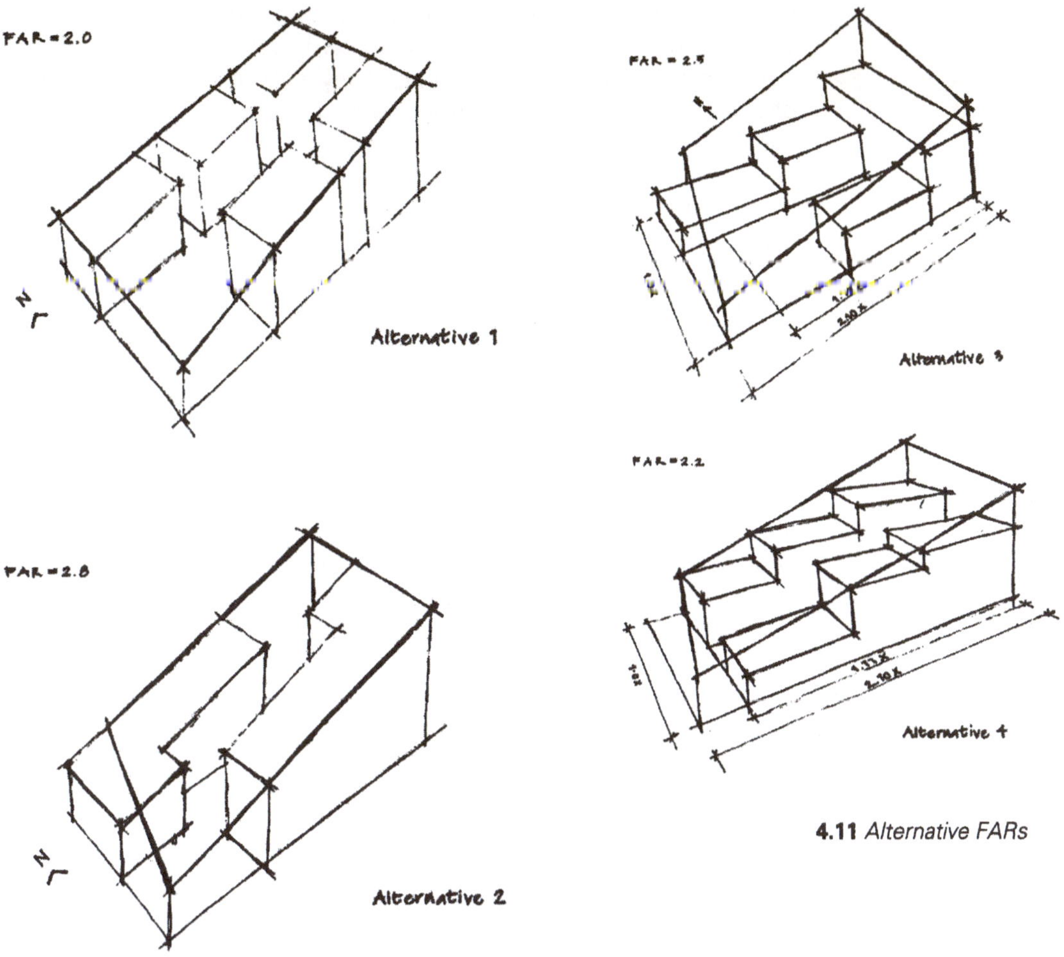

4.11 *Alternative FARs*

4.3 Promotion of urban wind flow

As discussed earlier, designing for wind penetration is difficult at best in the tropics. With low levels of macro wind flow and drastic seasonal changes in direction, designing for wind movement in tropical areas poses special problems to urban designers.

Urban wind flow

The urban boundary layer vs. canopy layer distinction introduced by Oke (1976) and detailed in Chapter 2 (see especially Figure 2.4) is crucial to the understanding of the urban wind flow patterns. Wind flows in the urban boundary layer (UBL) could be considered as the "undisturbed" wind (i.e., wind not influenced by urbanization). It is generated by the differences in atmospheric pressure caused by uneven solar radiation and resulting air temperature and density variations. This flow is modified by topography as well as by the presence of waterbodies. This "undisturbed" wind is called the "gradient wind" and its velocity is the "gradient velocity." Usually, the "gradient wind" flows at a height of several hundred meters above ground.

However, near the ground the wind experiences friction. Its speed is retarded drastically and its turbulence (i.e. speed variation) increases even within a short distance. Both parameters of wind (speed and turbulence) are affected by the topography as well as by the urban form. Elements that retard wind flow in urban areas are called the urban "roughness elements."

The urban roughness elements are mainly buildings that are rigid sharp-angled bodies distributed along streets. This causes the wind speeds to be retarded, but at the same time the turbulence of the flow is greatly enhanced. The specific design details of the buildings and the streets, the height of the buildings relative to one another, and the orientation of the buildings with respect to the wind, greatly affect the actual wind flow within the urban canopy layer.

The main design elements that modify wind flows within the urban canopy layer are:

- the overall density of the urban area;
- size and height of individual buildings;
- existence of high-rise buildings;
- street orientation;
- availability, size and distribution of open areas and green belts (Givoni, 1989).

Although several mathematical models describing urban wind flows within the canopy layer exist, two types of urban wind laws are more commonly used. The first type is called the "logarithmic" wind model. It is not only complicated and idiosyncratic but also useless in calculating wind flow within the canopy layer. The second type of wind model is referred to as the "power wind law." It is not only simpler but the data required to calculate urban wind flows are readily available.

We give below a "power wind law" developed by Davenport (1960):

$$\frac{V_Z}{V_G} = \left(\frac{Z}{Z_G}\right)^{\rho} \qquad (8)$$

Where,
V_Z = wind speed at height z;
V_G = height where the gradient wind starts ("gradient velocity");
Z = height for which wind speed V_Z is computed;
Z_G = height at which the "gradient velocity" (V_G) is first observed;

Table 4.4 Commonly suggested values of Z_G and ρ

Terrain condition	*Davenport (1960)*		*Poreh and Paciuk (1980)*		*Givoni (1989)*		*Chandra (1989)*	
	Z_G	ρ	Z_G	ρ	Z_G	ρ	Z_G	ρ
Ocean or >5 km expanse of water	–	–	–	–	–	–	250	0.10
Open flat country, flat coastal belts, desert	270	0.16	300	0.15	300	0.16	300	0.15
Low wooded areas, sparse trees, low-density rural sites, small towns	390	0.28	400	0.20	400	0.20	340	0.20
Wooded land with high trees, suburbs, urban meteorological stations	–	–	–	–	400	0.25	370	0.25
Urban areas with medium-to-high density, height of buildings 10 m	–	–	400	0.20	400	0.30	420	0.35
Centers of cities, buildings of more than 10 storeys	510	0.40	–	–	500	0.40	–	–
Centers of large cities, building height >30 m (over 30 floors)	600	0.50	400	0.28	600	0.50	–	–

ρ = an empirical exponent, which depends on the surface roughness, stability and temperature gradient (Givoni, 1989).

Wind speeds recorded at typical meteorological stations (open country, atop a wind mast, usually at 100 m [328 ft] high) could normally be substituted for gradient velocity (V_G). As for Z_G and ρ, various authors suggest conflicting values. Table 4.4 presents some of the more commonly suggested values, according to different types of terrain.

Although "power wind law" is capable of predicting wind speeds at any height within the urban canopy layer, the quantification of wind speeds using this law may sometimes appear unrepresentative of the urban wind field. In fact, Givoni (1989) states that defining a representative and meaningful urban wind speed by a simple mathematical model is questionable. "Representative" wind speed make sense in an open area (or a meteorological station), but is nonsensical in the urban context. Every building, every tree and indeed every roughness element in the urban canopy layer influences the wind speed and makes predicted wind speeds quite meaningless. There is no "general" mathematical model predicting urban wind fields near ground (say, less than 2 m above ground). Caution must therefore be exercised when using the "power wind law," which at best could only be used at a height near the top of the urban canopy layer. The art of predicting wind speeds at the pedestrian level remains largely uncharted.

Designing for urban wind movement in the tropics

The primary task of designing for wind movement should begin at city level, particularly in its street layout pattern. Streets should be so aligned to reap the maximum benefit from macro-level wind directions. In the tropics, this would usually mean aligning the streets along the major monsoonal wind directions.

In the weak wind regime of the tropics, another possibility is to induce wind flow by the thermal differences that arise at the edges of waterbodies. Differences in the thermal properties of land and water generate sea/land breezes at day/night respectively. These wind-flow patterns could be effectively used by sensitive urban planning measures that promote deep wind penetration into cities.

In addition to the above, the promotion of ventilation in tropical areas may benefit from the following considerations (Givoni, 1994: 1048):

1. Location of a town within a region.
2. Density of the urban area.
3. Orientation and width of streets.
4. Heights and relative heights of buildings.

Although a city street layout parallel to the major wind directions will be beneficial at the urban scale, it might cause wind flow problems at a building scale. Givoni suggests that the tropical street layout be at an oblique angle (between 30 and 60 degrees) to the prevailing winds (Givoni, 1994: 1050).

Built densities and heights are other important variables in promoting ventilation in the urban tropics. Usually high-density zones are prone to poor ventilation regimes, and long walls of building fabric prevent deeper penetration wind. Givoni (1994: 1050) suggests that "an urban profile of variable building height, where buildings of different heights are placed next to each other, and when the long façades of the buildings are oblique to the wind enhances urban ventilation."

References

Carmona, L.S. de (1986). "Human comfort in the urban tropics," in *Urban Climatology and its Applications with Special Regard to Tropical Areas*, Proceedings of a Technical Conference Organized by the World Meteorological Organization, held in Mexico City, 26–30 November 1984, pp. 354–404.

Chandra, S. (1989). "Ventilative cooling," in J. Cook (ed.), *Passive Cooling*, Cambridge, Mass.: MIT Press, pp. 44–48.

Correa, C. (1989). *The New Landscape: Urbanization in the Third World*, London: Butterworth Architecture.

Davenport, A.G. (1960). *Wind Loads on Structures*, Technical Paper No. 88, Ottawa, Canada: National Research Council.

Emmanuel, R. (1993). "A hypothetical 'shadow umbrella' for thermal comfort enhancement in the Equatorial urban outdoors," *Architectural Science Review*, 36(4): 173–184.

Emmanuel, R. and N.G.R. Perera (2002). "The bio-climatic approach to urban public space: a thermal evaluation of 'streets' in Colombo," in W.K. Hirimburegama (ed.), *Proceedings of the 58th Annual Sessions of the Sri Lanka Association for the Advancement of Sciences (SLAAS)*, 2–7 December 2002.

Gehl, J. (1989). "A changing street life in a changing society," *Places*, 6(1): 8–17.

Givoni, B. (1989). *Urban Design in Different Climates*. World Meteorological Organization Technical Note, WMO/TD- 346, WCAP-10. Geneva: World Meteorological Organization.

Givoni, B. (1994). "Urban design for hot, humid regions," *Renewable Energy*, 5(II): 1047–1053.

Hardin, G. and J. Badden (1977). *Managing the Commons*, San Francisco, Calif.: W.H. Freeman and Co.

Havlick, S.W. (1983). "An account of walking as a way of life in three European cities," *Ekistics*, 302(50): 351–355.

Illich, I. (1982). *Gender*, New York: Pantheon Books.

Knowles, R. (1981). *Sun Rhythm Form*, Cambridge, Mass.: MIT Press.

March, L. and L. Martin (1972). *Urban Space and Structure*, Cambridge: Cambridge University Press.

Nikolopoulou, M., N. Baker, and K. Steemers (2001). "Thermal comfort in outdoor urban spaces: understanding the human parameter," *Solar Energy*, 70(3): 227–235.

Oke, T.R. (1976). "The distinction between canopy and boundary-layer heat islands," *Atmosphere*, 14(4): 268–277.

Poreh, M. and M. Paciuk (1980). *Criteria for Identifying Wind Problems During Initial Planning Stages*, Haifa, Israel: Building Research Station, Technion (in Hebrew).

Tay Kheng Soon (1993). "The tropical city concept," in W. Attoe and R. Emmanuel (eds), "Hot-Humid Places: Essays in Life and Architecture," pp. 118–144 (unpublished manuscript).

Watson, D. and K. Labs (1983). *Climatic Building Design: Energy-Efficient Building Principles and Practice*, New York: McGraw-Hill.

Wright, C.A. (1992). *Fast Wheels, Slow Traffic: Urban Transport Choices*, Philadelphia, Pa.: Temple University Press.

Applications

CHAPTER 5

Urban design strategies

We have now established that urbanization alters the microclimate and therefore the thermal comfort of urban dwellers, and that this effect is greatest in the urban tropics. We have also defined basic concepts that can help ameliorate the negative thermal comfort consequences of urbanization. In this chapter we will look at ways in which the conceptual idea of the "shadow umbrella" so far developed could be applied to concrete urban situations in the tropics. While the idea of the "shadow umbrella" addresses the need for solar radiation prevention, we will also look at urban strategies that will help augment the cooling potential of urban shading. These include promotion of urban wind movement and use of evaporative cooling in the outdoors.

Such an approach will be the only rational way to ensure that nature takes hold of the tropical cities. Tropical rain forests are vanishing rapidly. In Sri Lanka, parts of southern India and tropical Africa, rain forests have actually been wiped out. In the face of rapid urbanization, nature in cities perhaps offers the last hope for recovering some semblance of green cover in the region.

The primary reason for taking a stand towards this approach is that a climate-sensitive urban form is the best bet for ecologically sensible, microclimatically appropriate high-density growth in the tropics. It is also the only humane way of providing dwelling for tropical people at a higher density. Architecture today dictates that responses are made within the confines of nature. "Ecological problems are solved as by-products of doing this" (Gray, 1979: 235).

5.1 Fundamentals of environment-conscious design in the tropics

Climate-sensitive design is synonymous with energy-efficient design. As we saw in the first chapter, energy-efficient design leads to improvement in the environmental quality. Although this is true at the micro-scale (building level), greater improvements to the environmental quality occur when urban design strives to achieve energy efficiency at the macro (urban) scale. Achieving energy efficiency at the city scale would be greatly facilitated if we conceive the city as a living organism. This approach (the so-called urban metabolism concept) was first introduced by Wolman (1965). According to the principle of urban metabolism, the environmental quality improvement in urban areas rests on the *careful use and removal of energy and matter.* Application of the urban metabolism concept will achieve the following goals:

- energy-efficient urban layout;
- minimize input quantity and maximize output quality of urban water use;
- reduce the need for fossil-fuel-powered transportation.

In this chapter, we will present the basic principles that underlie each of these goals, while detailed design strategies are presented only for the achievement of the first two goals. Urban design strategies that lead to energy efficiency in transportation are explained in the following chapter.

Energy-efficient urban layout

Patterns of energy use differ among nations, yet certain commonalties exist in their urban energy needs. A comparison of energy-use patterns between a developed country like the US and a developing country like Sri Lanka shows that transportation and activities within buildings consume a considerable share of energy in both cases.

Urban design has an influence on energy needs for transportation and in building energy use. Table 5.1 shows that this influence is significant, irrespective of the socio-economic context of the country.

Energy consumption in the developing world has grown tremendously over the past years. A World Bank study reports that the developing world, which accounted for 20 percent of the total global consumption in 1970, increased its share to 33 percent in 1988 (Imran and Barnes, 1990). This trend is expected to grow at least for the first three decades of this century.

Table 5.1 Patterns of commercial energy consumption

Activity	*Energy consumption (% of total country needs)*		
	U.S.A.	*U.K.*	*Sri Lanka*
Industries	41.2	32.0	9.9
Transportation	21.0	18.0	16.4
Building energy needs	28.0	48.0	67.0
Agriculture / Other	7.7	2.0	6.7

Source: Emmanuel (1995)

Although many factors are cited as causes for the increase in consumption, it is generally agreed that urbanization in the region is the prime cause, particularly so in areas where increasing income has resulted in greater demand for transportation and home appliances – especially space-conditioning devices (see de Dear and Fountain, 1994).

While energy efficiency can be achieved at several scales, a useful starting point is the very form of the city. Overall form determinants need to be put in place right at the beginning if we are to regulate urban development, which is usually piecemeal, in a planned manner. In the context of the tropical climate, energy-efficient urban layout needs to address the following issues:

1. Site and building orientation for avoidance of sun.
2. Site and building orientation for encouraging air movement.
3. Shading device requirements for open spaces.
4. Landscaping requirements for heat-gain reduction.

At the same time, the efficiency of urban activities depends on the number of people served: in other words, density. And density is not necessarily proportional to height (March and Martin, 1972). Density is achieved by a reorganizing of street patterns and rethinking of the importance of the automobile to urban life.

Another strategy for dealing with density is to restrict plot sizes. In the Indian context, Correa (1989) suggested a medium size of 50–100 m^2 (540–1080 ft^2) in his new Bombay proposals and found that they were acceptable for both the poor and the affluent sections of the society.

Reduce the need for fossil-fuel transport

Urban design strategies that improve the transportation regime in cities are dealt with in great detail in the following chapter, but let us look at the rationale for considering transportation in a textbook such as this one. The greatest impediment to urban quality of life comes from automobiles. Table 5.1 points out that up to a fifth of the energy used in developed countries (and somewhat less in the developing world) is consumed by the transportation sector. Thus, energy efficiency in cities cannot be achieved if one ignores urban transportation. The most sensible approach will be the encouragement of the most energy-efficient modes of transportation. These include walking and bicycles, closely followed by public transportation (see, for example, Illich, 1974; Owens, 1986, 1987; Correa, 1989; Lowe, 1991).

The problem with walking and riding bicycles in the tropics is that the high air temperature, coupled with high relative humidity, makes these activities enervating (Ondaatje, 1979). To encourage these modes, therefore, shaded pathways exclusively designated for pedestrians and bicycles need to be provided. These actions will invariably impinge upon the climatic quality of the urban areas. Thus a comprehensive treatment of climate-conscious urban design cannot afford to ignore urban transportation.

Minimize input quantity and maximize output quality of urban water

The urban water cycle is both large in volume (due to increases in supply and waste – see Figure 5.1) and poor in quality (due to pollutants, heavy runoff and turbidity). The water balance of an urban system is given in Figure 5.2 and by the following model:

$$p + F + I = E + \Delta r + \Delta S + \Delta A \qquad (1)$$

where p is precipitation, F is automobile and industrial combustion release, and I is water supply (i.e. the "supply" side); E is evaporation from urban system, Δr is water output (runoff and refuse), ΔS is storage in the fabric, and ΔA is net horizontal movement (advecture) (i.e. hydrological sinks).

While urban design may not have much influence on the supply side of urban hydrological balance, it certainly can influence the relative partitioning of output water into hydrological sinks. Generally speaking, a greener city has more possibilities for increased latent heat partitioning (and therefore less sensible heat outflow, less urban heat island effect), lower thermal inertia in the urban fabric (thus lower heat storage), and slower – but higher quality – *runoff* from the urban system.

Enhancement of evaporation (E) involves increasing green surfaces in the urban area as well as impoundment of rainwater. Storage (ΔS) can also be improved by encouraging porous urban surfaces. The quantity of net water output (Δr) can be reduced by impoundment as well as by using waste water as a useful by-product of the urban system. This in turn can improve the quality of (Δr).

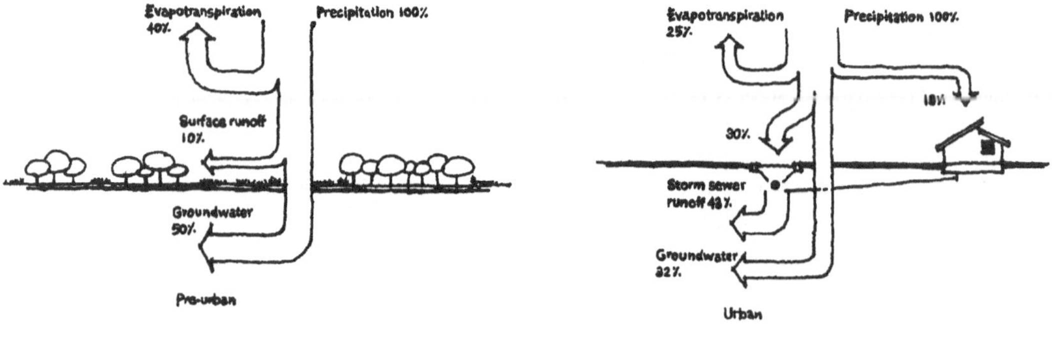

5.1 *Urban/rural hydrological differences.* Source: Hough (1984)

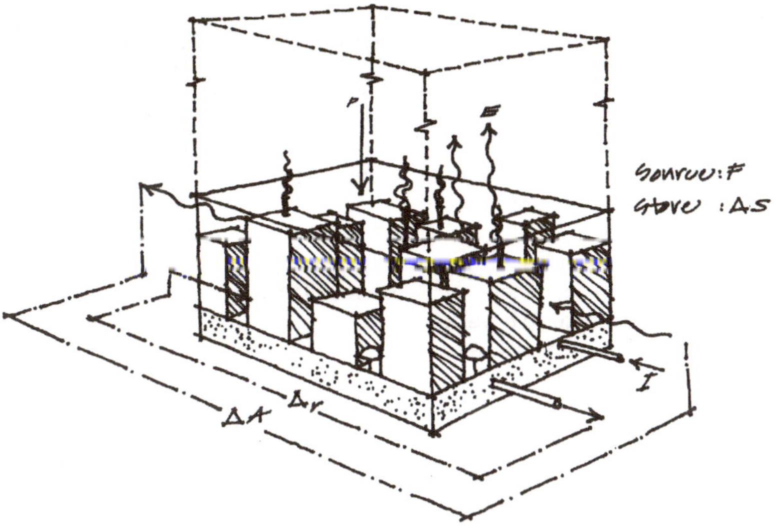

5.2 *Water balance in cities.*
Source: Oke (1987)

The general principle of urban hydrological management is best captured by the phrase, "reduce need and improve output quality." It would translate into at least four design goals:

- waste as resource;
- land as a filter;
- retention ponds and lakes;
- enhancement of green surfaces.

5.2 Tools for enhancing urban environmental quality

Urban design protocol offers environment-conscious designers at least three tools for the realization of the goals of thermal comfort improvement (energy efficiency), transport reduction and water quality improvement (Table 5.2).

- zoning laws;
- building laws;
- landscape control.

Some attempts at utilizing these tools for the purposes of energy and transportation need reduction have already been made (see Emmanuel, 1995). Although these attempts are from the temperate climate cities they offer possible models for tropical cities (Table 5.3).

Table 5.2 Design tools for urban physical quality enhancement

Design tool	*Possible form*	*Design goal*	*Environmental effects*
Zoning law	Subdivision size regulation, built density control, land-use control, street width and type control, specifications for bicycle lanes and pedestrian paths, promenade and neighborhood squares regulations, waterfront development controls	Energy efficiency, transportation reduction	Air quality, climate quality improvements
Building regulations	Building and site orientation guidelines, building form guidelines, building envelope control, arcade development guidelines, building height limits, shading requirements, energy audits	Energy efficiency	Climate quality improvements
Landscape control	Type and density of green space guidelines, water impoundment requirements, hedge and fence controls, public green space guidelines	Energy efficiency	Climate quality, water quality improvements

Table 5.3 Some examples of existing guidelines for energy-efficient urban design

Type	*Location*	*Description*
Cooling need reduction	Port Arthur, Texas	Subdivision regulation that requires passive solar orientation of sites to reduce solar exposure of buildings and windows in summer
	Sacramento, California	Street and lot orientation encouraging buildings to be oriented for passive heating and cooling
	Dade County, Florida	Site plan review criteria incorporating requirements for energy-efficient orientation
	Boulder, Colorado	Incentives to encourage energy efficient site layout
	Douglas County, Kansas	Permission to build "earth sheltered" buildings
	King County, Washington	Permission to develop "townhouse"-type dwellings of high density
	Davis, California	Flexible siting of hedges and fences so as not to interfere with heat collection
	Davis, California	Permission to use shading devices even within open spaces
	Davis, California	50 percent landscaping requirement for parking lots
Transportation need reduction	Boulder, Colorado	New development to be located near existing facilities
	Windsor, Connecticut	Encourage conversion of one- or two-family houses into multi-family, multi-use buildings
	Davis, California	Encourage home occupations

Source: Adapted from Emmanuel (1995)

5.3 Design strategies for the urban tropics

Design strategies for the enhancement of the urban physical environmental quality are specific suggestions utilizing one or more of the above-mentioned design tools. In order to achieve the design goals of energy efficiency, transportation reduction and water quality improvement, design strategies could take one of the following forms:

- building form guidelines;
- activity pattern controls;
- control of relationship to natural features (Emmanuel, 1994).

The primary controlling factor in all these strategies is the "shadow umbrella" developed in Chapter 4. While shading promotion is the primary concern, they augment the cooling potential of shading by promoting urban ventilation and evaporative cooling.

Building form

Courtyard forms

Courtyards (Figure 5.3) have had long architectural traditions. Though originating in hot-dry areas like in the Middle East, European influence in the hot-humid region has produced adapted courts for various purposes, primarily dwelling.

5.3 *Courtyard-shaped forms*

While the courts in hot, dry zones have a cooling function (drawing hot air towards the cool court), the possibility of creating a shaded "common" in the middle of the house is an appealing proposition for tropical dwellers. However, care must be taken not to jeopardize thermal comfort in the built spaces. There is proof that wrongly oriented and proportioned courtyard buildings are detrimental to thermal comfort in weak-winded areas like the equatorial tropics.

Therefore, where private "commons" are desired, let the buildings enclose a court. The buildings themselves must be minimal, as thin as possible, the emphasis being on the shading they create upon the court. The height of building will be the critical factor, essentially following the shadow profile mandated by the "shadow umbrella." A three-sided court with no buildings on the west will be the ideal, the west being the most difficult orientation to shade the low latitudes.

Orientation

The high altitude of the sun during most of the year in the hot-humid regions dictates that west and south façades are the most difficult to shade. Thus site-layout and building-orientation incentives can be built into the building ordinances of tropical cities to encourage north/east orientations.

Activity pattern controls

> Here the colonnade is not just a (heavily coded) screen through which you see the main building, but a perfectly pleasant spot to saunter around for much of the day. And the monumental Hindu temples of south India – at Madurai, at Tanjore, at Srirangam – are experienced not just as a collection of gopurams and shrines, but as a pedestrian path (a pilgrimage!) through the sacred spaces that lie between. In fact, this open-to-sky processional movement is of the utmost religious and symbolic significance. It is found throughout the warm regions of the earth, from the Sun Temples of Mexico . . . to the Temples of Bali . . . (Correa, 1983: 43)

In the warm regions of the world, moving is enervating. At the same time, "if you stood still you were invaded" (Ondaatje, 1979: 189). There is a slow determined fatefulness about moving in the tropics. Sweat pours down your back at the suggestion of movement. Shading is therefore the key. But thermal comfort cannot come from shading alone when one moves in the urban outdoors, for one is not staying long enough at a place to achieve a stable thermal equilibrium with the environment. Air movement can therefore be most welcome when moving.

Shopping streets

Certain types of specialized vending tend to be located in large groups. Thus, every town has its own streets of jewelers, textile businesses, stationers, etc. There is an inherent sense of satisfaction and a feeling of choice that arises from shopping in such specialized streets. The movement from one vendor to another is part of the bargaining process that shoppers undertake almost ritually. And this movement must be predominantly pedestrian. While the shoppers need access to these shopping streets, the actual activity of shopping itself does not benefit from traffic (Alexander *et al.*, 1977: 175). Therefore, the need is to create a built form that is of a humane scale, close enough to cast a shadow on the street itself, offering protection from the elements and discouraging motorized traffic.

Therefore, create canyon-like shopping streets in large neighborhoods. The street itself should preferably be north/south running, with buildings on the west being slightly taller than those on the east (Figure 5.4). Line both façades of street-facing buildings with arcades so as to provide adequate protection from rain and sun. If north/south running streets are not available, create narrow shopping culs-de-sac off major traffic arteries, thus ensuring adequate protection from traffic, while maintaining accessibility.

Gathering places

While all climates need gathering spaces, the particular need in tropical climates is for negative ionization, a potential brought out by carefully designed outdoor environment. The presence of water has no rival in the tropics to match its revitalizing negative ionization effect. Thus, one could see thousands of city dwellers on Colombo's waterfront every evening. The same is the case on Chennai's Marina beach, the second largest in the world. Rio de Janeiro's waterfront promenades of course are legendary. People go there every day (except when it rains continu-

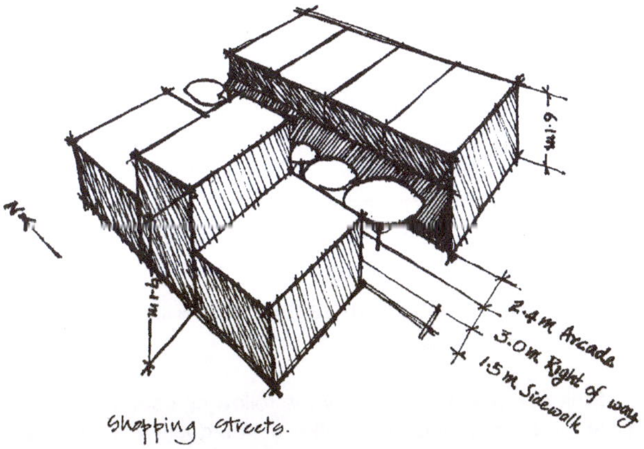

5.4 *Shopping streets*

5.5 *Gathering places*

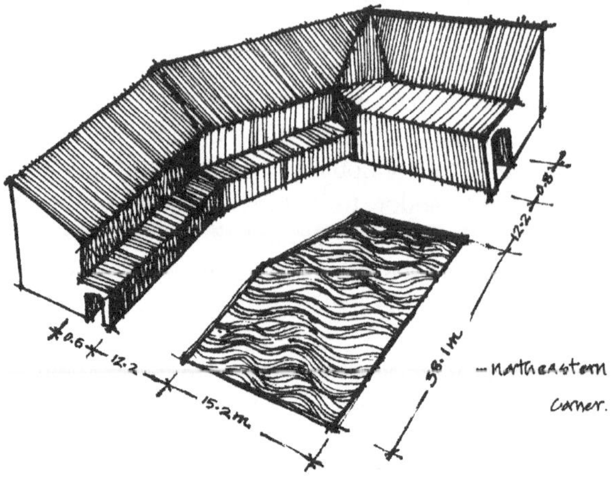

ously), even though some of these environments are not thoughtfully developed. Of the four types of urban-space needs in the tropics identified by Correa (1989) and referred to in the previous chapter, three are gathering spaces.

The tropics are blessed with highly usable outdoors.[1] The urban design issue therefore is one of thoughtful manipulation of outdoor design variables like built massing, orientation, landscaping, street furniture, etc., for gathering.

Therefore, surround urban waterbodies with a built form edged by arcades (Figure 5.5). Locate public activities along the arcade, overflowing into the commons. If north/east-facing promenades are desired, locate the buildings in an "L" shape along the southern and western edges of water. If south/west facing arcades are desired, locate buildings along the northern edge of water, in a staggered manner. The commons in the latter case needs to be wider. Trees need to be planted so as to provide intermittent shading to the commons. Dissuade all activities that do not directly benefit from the water's edge from locating on urban waterfronts.

Provisions for evening life

Evenings are the tropics' winter (Rao, 1981). The places deserted in the heat of the day are slowly reoccupied by throngs of people as the searing afternoon sun bids farewell and darkness envelopes the land. Thus, city centers in southern Indian small towns, and beach resorts from Sri Lanka to Brazil, begin to come alive at dusk with evening bazaars and pedestrians.

Street corners are especially attractive to be incorporated into the

[1] The usability of the outdoors is defined by the following coefficient (Correa, 1989): usability coefficient = % of activities that can occur outdoors × % dry months.

evening life, particularly if they intersect northerly and easterly orientation.

Pedestrian paths and nodes

There are two kinds of traffic: transport which is motorized, capital-intensive and therefore scarce by definition, and transit which is self-powered and therefore labor-intensive. In the resource-starved hot-humid regions it is clear that traffic must be biased towards transit (see Illich, 1974). It is the most energy-saving and ecologically sound thing to do in the region.

And pedestrianizing is not returning to the past either. Even in a highly motorized city like New York, more than 70 percent of all downtown rush-hour traffic is on foot (Wright, 1992). Even in a highly motorized country like the US, an average person spends 1,600 hours a year with his car (the actual time spent at the wheel, finding a parking space and walking to and from it, time spent in repairs, etc.) to travel 16,000 kilometers, an average speed of 10 km.p.h. – hardly faster than walking (Illich, 1974).

The problem of pedestrianizing cities lies in the connections provided. It is not only the provision of a safe lane, but also connecting activities that people most frequently use, is necessary. It must also be remembered that most people have an upper limit on the distances they prefer to walk. Some kind of mixing of pedestrian and motorized modes is therefore necessary.

It goes without saying that the whole path must be climatically

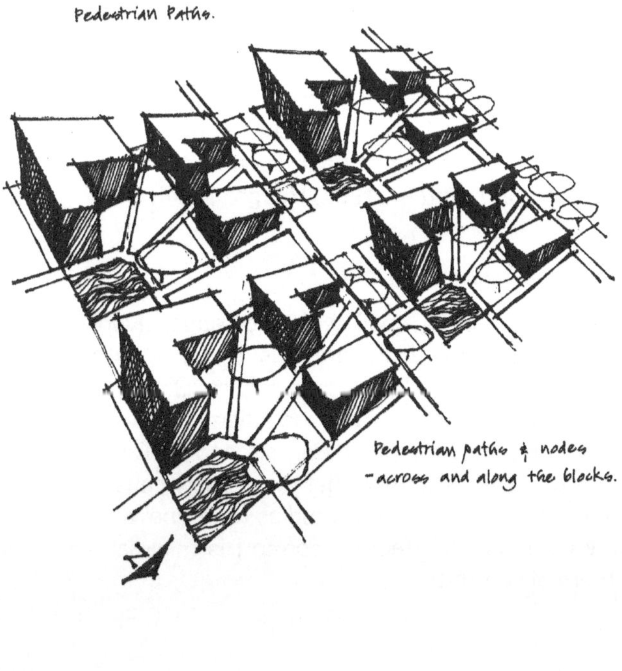

5.6 *Pedestrian paths and nodes*

appropriate for movement in the tropical regions. This would largely mean ample shading.

Therefore, establish a network of pedestrian links independent of the city street network (Figure 5.6). Paths within city blocks could be the primary carriers of cool air (say from a pond). Wherever the pedestrian path crosses the vehicular path, establish a place to pause. At places, let the pedestrian path merge with the street network, especially if the street is north/south running. Let such streets be very narrow, buildings on both sides providing shade to the entire street. Line the rest of the path with trees, or, alternatively, cover it with a light roof.

Network for cars

The car is a necessary part of modern urban life. Just as pedestrians need cars at some point, cars also need people. While too much merger is dangerous for the pedestrians, total separation of the two results in dead spaces.

Therefore, reorganize the city street network by converting the east/west running streets into vehicular paths and north/south running ones into pedestrian paths. The former can be wide and lined with arcades, and not entirely in the shade. (It is easier to shade south- or north-facing buildings that line up such streets.) The other network for pedestrians will be perpendicular to the vehicular paths and narrow. Where these modes meet, establish an activity node where people can linger.

Relationship to natural features: landscape controls

The cooling potentials of natural elements like water and vegetation in a hot-humid place should be well utilized. Yet urban environments also are harmful to the well-being of vegetation and waterbodies (see, for example, Sukopp and Werner, 1982; Hough, 1984).

Relationship to waterbodies (Figure 5.7)

The potential of water in cooling is substantial even in the humid regions (Cook, 1989). However, care must be taken not to overload the already heavy relative humidities of the tropics (Givoni, 1992).

Fanger's (1970) thermal comfort equation shows that high humidities can be compensated by lower air and radiant temperatures. Although it is generally assumed that the primary cause of thermal discomfort in the tropics is humidity, studies have shown that at lower air temperatures humidity's effect on comfort is minimal (see Chapter 3 on urban thermal comfort).

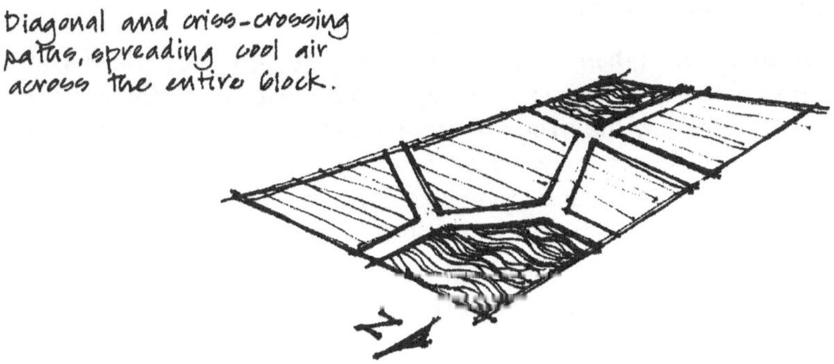

5.7 *Relationship to waterbodies*

The design strategy in incorporating water into urban design in the humid regions is one of careful provision of just adequate water for evaporation. Furthermore, if waterbodies can be located to take advantage of the general air movement patterns in the region, the air itself can be "conditioned," thus increasing its *potential* for cooling (Figure 5.8).

Therefore, locate waterbodies in every neighborhood at the northeastern/southwestern corner of city blocks. Let some major traffic routes (pedestrian paths if within a city block) connect these pools of water diagonally. At larger scales, rearrange the street network to take advantage of air movement thus created by such diagonal paths (i.e. perpendicular to the diagonal paths). Plant trees around waterbodies to increase the cooling potential of air.

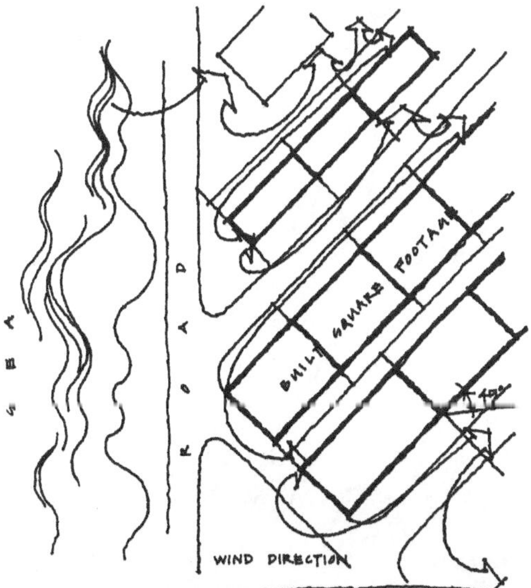

5.8 *Orientation strategies*

Contiguous green areas

> Much of the urban open space is fragmented into . . . lawns and small gardens. Although in the aggregate these are highly desirable, they affect only the microclimate wherever they are located. Large, coherent park areas, however, have quite a measurable influence. In warm summer nights, the parks are cool . . . up to 3°C, and 1–2°C at the time of maximum temperature. (Landsberg, 1981: 232)

Instead of opening up large parts of cities to greenery, contiguous greens can be provided that take into account the general wind patterns of a city (Figure 5.9). Thus the cooling potential of air can be enhanced, the ability of greens to reduce flash floods increased, and the spread of evapotranspiratory agents can be made more even. The size of greens, however, is a critical factor.

Another important factor is the type of greens provided. While most green areas in cities are park-like (single or groups of trees amidst lawns), the climate-enhancing potential of green areas can be increased by the provision of urban forests, where wild growth of trees is encouraged.

Therefore, specify not only the size but also the orientation of open spaces in city blocks. Connect as many such open spaces as possible by public parks. Some of the parks can be allowed to grow wildly. Make every effort to connect green areas in a diagonal manner (northeast–southwest). Combine such connected greens with the urban pedestrian network.

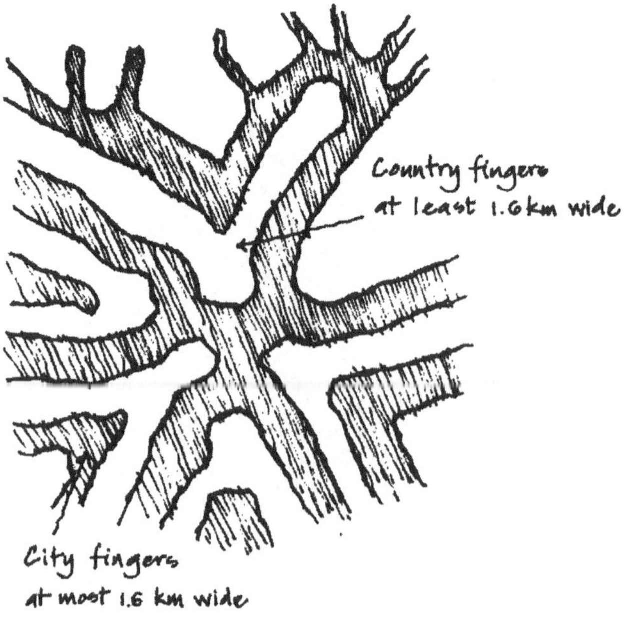

5.9 *Contiguous green areas.*
Source: Alexander *et al.* (1977)

Collection of rainwater

It rains for an average of 150–180 days a year in the tropics, particularly in the equatorial region (see, for example, Rao, 1981; Nieuwolt, 1981). Periodic flooding of urban streets is a frequent phenomenon in tropical cities. It is a pity that such abundance of water is not utilized for urban climate amelioration.

It is well known that urbanization increases the chances of flash flooding (see, for example, Chandler, 1976; Landsberg, 1981). There are fewer possibilities for rainwater to linger in cities. Our ideas of "keeping dry feet" often clash with urban hydrology. Furthermore, the rapid encroachment upon urban wetland fringes makes cities even more vulnerable to flooding.

Therefore, collect rainwater runoff from roofs, etc., to form shallow pools of water on the western edge of city blocks. Since these edges are hard to shade and therefore have to be left bare, the presence of water can be a welcome relief.

Topographical relationships

> They spent most of the year in Colombo and during the hot months of April and May moved to Nuwara Eliya (located in the central hills of Sri Lanka at an elevation of 1,800 m above sea level). In various family journals there are references made to the time spent "up-country" away from the lowland heat. Cars would leave Colombo . . . radiators steaming as they wound their way up into the mountains. Books and sweaters and golf clubs and rifles were packed into trunks, children were taken out of school, dogs were bathed and made ready for the drive . . . (Up country) was a different world. One did not sweat there and only those who had asthma tried to avoid vacations (Ondaatje, 1979:39).

The effect of elevation change on tropical climate is remarkable; so remarkable that upland areas are designated as separate climatic zones. Even a slight change in grade is instantly noticed by dwellers in these regions who constantly bathe in sweat.

Therefore, if new urban development is planned, give preference to north/east-sloping sites. Avoid west-sloping sites. If unavoidable, permit only very short buildings (maximum height: single-storey) on west slopes. If the site is north-sloping, the whole "shadow-umbrella" can be raised according to the slope. If east-sloping, the "shadow umbrella" will be taller only on the east. If west-sloping, the built massing will have to be lowered proportionately.

5.4 Conclusions

Urbanization has a dynamic relationship with the physical environment. While urban growth affects the physical environment (usually negatively), urban environmental changes also affect the quality of life in these areas. The latter elicits biochemical, epidemiological and psychological responses in the urban dwellers. The debate over the relative importance of social vs. physical environmental factors on these responses is not yet settled.

The two-way dynamism of environment–urban relationship was discussed in terms of three key physical environmental attributes: air, water, and climate. It was argued that enhancing the urban quality of life in the physical sense involves at least three concepts: energy-efficient urban form, transport-need reduction, and water-quality improvement. Three design tools available for the urban designers in the management of urban environmental quality were also enumerated. These were: zoning laws, building and site regulations, and landscape controls. Specific design guidelines in the context of a hot-humid climate were also proposed.

However, these outlines are hypothetical in that there is little hard evidence for climatic, air-quality and water-quality changes in the hot-humid urban areas, particularly the developing regions. As pointed out earlier, tropical urban climate is only beginning to be studied now. Thermal comfort research on this region is yet to produce an indigenous quantitative index. The monitoring of air- and water-quality changes in the region is at most rudimentary (*Urban Air Pollution in Megacities of the World,* 1992). Efforts at monitoring urban environmental changes and quantifying them must precede attempts at mitigating them.

References

Alexander, C.A. *et al.* (1977). *A Pattern Language,* New York: Oxford University Press.

Cook, J. (ed.) (1989). *Passive Cooling,* Cambridge, Mass.: MIT Press.

Correa, C. (1983). "A place in the sun," *Places,* 1(1): 40–49.

Correa, C. (1989). *The New Landscape: Urbanization in the Third World,* London: Butterworth Architecture.

de Dear, R.J. and M.E. Fountain (1994). "Field experiments on occupant comfort and office thermal environments in a hot-humid climate," *ASHRAE Transactions,* 100(2): 457–475.

Emmanuel, R. (1994). "Recovering the commons: a new approach to climate-conscious urban design in the equatorial tropics," Paper presented at the 12th Annual ACSA Technology Conference, in Jong-Jin Kim and Colin Clipson (eds), *Design and Technological Innovation for the Environment,* Washington, D.C.: Association of Collegiate Schools of Architecture (ACSA), pp. 124–128.

Emmanuel, R. (1995). "Energy-conscious urban design guidelines for warm-humid cities: strategies for Colombo, Sri Lanka," *Journal of Architectural and Planning Research*, 12(1): 58–75.

Fanger, P.O. (1970). *Thermal Comfort: Analysis and Applications in Environmental Engineering*, New York: McGraw-Hill.

Givoni, B. (1992). "Climatic aspects of urban design in tropical regions," *Atmospheric Environment*, 26B(3): 397–406.

Gray, C.B. (1979). "Man and architecture: the highmindedness of Paolo Soleri," *Urban Ecology*, 4(3): 235–244.

Hough, M. (1984). *City Form and Natural Processes: Towards a New Urban Vernacular*, London: Croom Helm.

Illich, I. (1974). *Energy and Equity*, London: Calder and Boyars.

Imran, M. and P. Barnes (1990). *Energy Demand in the Developing Countries: Prospects for the Future*, World Bank Staff Commodity Working Paper No. 23, Washington, D.C.: The World Bank.

Landsberg, H.E. (1981). *The Urban Climate*, New York: Academic Press.

Lowe, M. (1991). "Bicycle, vehicle for a small planet," *Ekistics*, 58(348/349): 179–194.

March, L. and L. Martin (1972). *Urban Space and Structure*, Cambridge: Cambridge University Press.

Nieuwolt, S. (1981). "The climate of continental southeast Asia," in K. Takahashi *et al.* (eds), *Climates of Southern and Western Asia*, World Surveys of Climatology, Vol. 9, Amsterdam: Elsevier Scientific Publishing, pp. 1–66.

Ondaatje, M. (1979). *Running in the Family*, New York: W.W. Norton.

Owens, S. (1986). *Energy, Planning and Urban Built Form*, London: Pion.

Owens, S. (1987). "The urban future: does energy really matter?," in D. Hawkes *et al.* (eds), *Energy and Urban Form*, papers presented at the International Seminar on Urban Built Form and Energy Analysis, held at Darwin College, Cambridge, 26–27 June 1986. London: Martin Centre for Urban Studies, pp. 169–186.

Rao, Y.P. (1981). "The climate of the Indian sub-continent," in K. Takahashi *et al.* (eds), *Climates of Southern and Western Asia. World Survey of Climatology*, Vol. 9, Amsterdam: Elsevier Scientific Publishing, pp. 67–82.

Sukopp, H. and P. Werner (1982). *Nature in Cities*, Nature and Environment Series No. 28, Strasbourg: Council of Europe Publication.

Urban Air Pollution in Megacities of the World (1992). Cambridge, Mass: Blackwell.

Wolman, A. (1965). Quoted by R. Stren, R. White, and J. Whitney (eds) (1992) *Sustainable Cities: Urbanization and the Environment in International Perspective*, Boulder, Colo.: Westview Press.

Wright, C.A. (1992). *Fast Wheels, Slow Traffic: Urban Transport Choices*, Philadelphia, Pa.: Temple University Press.

Climate-sensitive urban transportation for the tropics

CHAPTER 6

Energy-efficient urban design is not complete without consideration of urban transportation. City form, its spatial features, density, activity mix and its historical circumstances all contribute to the traffic ills of a given city, but climatic suitability of transport facilities underpins the energy requirements for urban transportation. Very real possibilities exist in the realm of energy-efficient *and* thermally comfortable urban transportation, especially in tropical cities. Yet current development ethos ignores these possibilities.

While the wastage of energy is an obvious consequence of ill-planned urban transportation, what concerns us most in these pages is the effect of such actions on the climatic health of tropical cities. This is primarily due to two reasons. On the one hand, wasted heat from transportation is a leading cause of UHIs even in the tropics. On the other, a workable regime of energy-efficient transportation in the tropics cannot be implemented without due consideration to the urban climate.

The benefits of climate-sensitive urban transportation are manifold. Primarily, we can list five attributes:

- improvement to the air quality;
- increased energy efficiency at the city scale;
- favorable impact on UHIs;
- enlivening of the "commons";
- favorable to the urban poor.

It is clear from Chapter 1 that urban transportation is the main source of air pollution (see Table 1.1) as well as of anthropogenic heat input to the urban thermal system. The way urban activities are organized influences the way energy is spent in commuting between them. The greater the diversity of activity, the lesser the need for mechanized transportation. The solution lies in providing the right mix of activities

Table 6.1 Annual travel distances by car in US households

Purpose	*Vehicle-km per household*		*Change 1969–1990(%)*
	1969	*1990*	
Work commute	6,730	7,808	16
Shopping	1,495	2,804	88
Personal	2,043	4,850	137
Social and recreation	6,587	6,533	−1
Total	16,855	21,995	30.5

Source: Lowe (1994: 83, table 5.1)

that will reduce the energy spent on transportation. The goal is not to smooth traffic flow, but rather to reduce it (Table 6.1). As Correa says, "ease of transportation is not an end in itself: Journeys always multiply to a point of clogging" (Correa 1989: 73).

Energy-efficient urban design in the tropical context would entail at least the following:

- incentives for energy-efficient site development;
- greater diversity of land-use development;
- encouragement of efficient transport forms.

Mechanically biased transportation strategies are not only energy wasteful but also biased against the poor. This has implications not only for energy conservation but also for environmental sustainability. In terms of energy use the Third World's consumption, though small now, is expected to overshadow that of the First World sometime during the early years of the twenty-first century. Rapid population growth, increasing affluence (particularly in the urban areas), and structural problems are the causes for this expansion. Most energy consumed in the Third World is imported, and the share of oil imports on national income is significant (Table 6.2). Most of this commercial energy is consumed by one sector: transportation. Needless to say, urban transportation plays an important part in this conspicuous consumption.

Mechanically biased transportation cannot be successful, particularly in the Third World. The reasons, however, are not just the energy costs, as indicated above, but also the environmental and ecological costs. By "environment" what is meant here is not just the physical environment but also non-physical entities like community, livelihood, freedom, etc. The hypothesis here is that urban transportation strategies that will work for the resource-starved Third World lie within the paradigm of biased innovations that strengthen the public realm of streets and empower individual freedom of movement at minimal cost both to oneself and society.

Table 6.2 Energy dynamics in selected tropical / partially tropical countries

Country	*1985*			*1990*		
	1	*2*	*3*	*1*	*2*	*3*
Bangladesh	51.9	91.5	5.4	43.7	111.2	5.9
Nepal	65.1	73.0	2.6	67.8	128.9	4.0
Pakistan	21.1	45.5	3.4	–	–	–
Sri Lanka	85.9	39.0	8.7	84.3	47.6	10.4
Low-Income	50.7	46.8	3.6	46.0	28.5	2.3
Papua New Guinea	91.5	18.8	10.2	87.9	23.0	11.9
Philippines	54.3	34.1	6.8	34.7	29.9	5.5
Thailand	66.3	49.2	10.3	59.3	52.2	10.6
Lower Middle-Income	61.5	39.0	8.3	48.3	39.6	8.1
Fiji	91.1	43.0	6.0	85.4	58.3	6.5
Hong Kong	100.0	13.5	10.4	100.0	16.3	14.3
South Korea	71.6	28.7	10.8	68.9	17.9	9.7
Singapore	100.0	6.3	10.7	100.0	6.2	11.9
High/Upper Middle-Income	77.1	21.2	12.5	77.9	21.4	15.2

Source: Emmanuel (2001)

Notes: 1 = Share of energy imports in commercial energy consumption (%)
2 = Energy imports as a percentage of export
3 = Energy imports as a percentage of GNP

A greater degree of congestion occurs as more and more rights of way are allocated to fewer and fewer individuals. The loss of public spaces pushes those who cannot afford the "private" into the exterior regions of the city (as opposed to the American cities where the more disadvantaged are in the central city areas). Thus, a specific way of ensuring the rights of as many people as possible is to preserve and enhance the public areas of a city: the commons or the public realm.

Choosing one's own mode of transport is a fundamental tenet of human rights. Current transport strategies, however, are nothing but a use of subtle force, compelling one to choose between "mass transit," that transports people like cattle, and private transport that is beyond the reach of most people in the Third World and not sustainable even by those who can afford it. Transportation strategies that suit the Third World cities well, it is argued here, must therefore encourage walking and cycling as modes of transit.

In the following pages, we will analyze the problem of Third World urban transportation and offer strategies for its improvements within the framework of urban design. Urban transportation is a function of the overall city form, and lasting structural change can only come from altering the direction of development of the urban form. It is recognized that urban design is a piecemeal activity, growing very slowly at the initiative of many different people acting in space and time. Design solutions are therefore presented as strategies that can be implemented individually yet direct the growth of urban form towards a sustainable form.

The following discussion is in two parts. In the first we look at the effect of city form upon urban transportation. A historical overview of the issues from around the world, with special emphasis on Third World cities, is presented. The current situation with respect to automobile use in Third World cities is also discussed. In the second part an alternative strategy for the provision of urban transportation is presented with the aim of achieving equity and sustainability. Increasing technocratization of modern life leads to more lasting and deeper social divisions, and policy planners on both sides of the ideological divide – left and right – are equally oblivious to this "modernization of poverty", to borrow a term from Illich (1974). The fundamental flaw in both world-views is that they tend to regard transportation as a mere economic activity. This can be seen in both the developed and the underdeveloped world: people opting for private modes of transport in the former and forced into the "public" mode in the latter. But the need to move about is not equal to transportation.

6.1 Urban form and transport

From the point of view of urban transportation, a great many of the world's cities could be characterized by three models:

- concentric-zone model or monocentric city;
- multiple nuclei model or polycentric city;
- sector model (see Pederson, 1980: 22–25).

The first model, the *concentric zone*, is the simplest form of urban development in terms of urban transportation (Figure 6.1). At the core of the city lies the highest-priced location: the Central Business District (CBD). Of course, it was not always the case that this area was the highest-priced location in a city's historical evolution, or called the CBD at all times.

Initially, the core of the city was all that there was when the city was formed. Thus the fort areas of Colombo and Mumbai, Inner London and a great many historical cities started as small nuclei that were pedestrian in character, densely populated and usually overcrowded.

When overcrowding and transport ills began to appear in the concentric-zone model, cities began to encourage several centers of influence. This led to the formation of the *multiple nuclei model* (Figure 6.2). Los Angeles was one of the early examples (1930s) of such development.

In this model, the importance of the CBD is diminished (there may not even be a CBD at all). Many city centers (nuclei) are present and

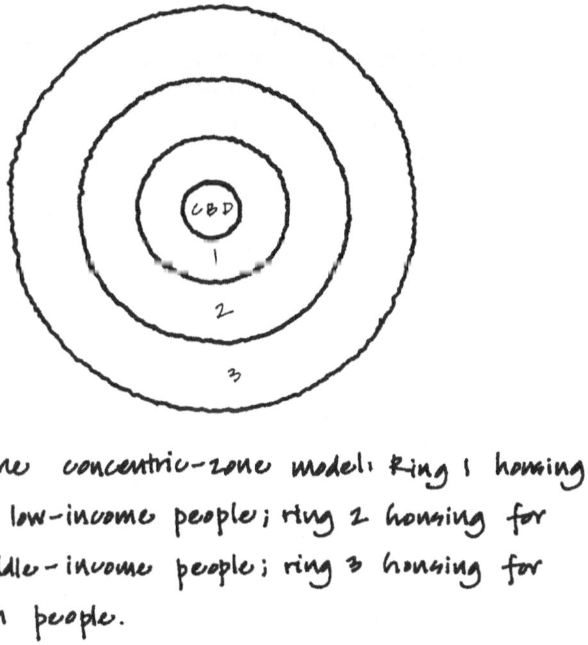

6.1 *Concentric city.* Source: Adapted from Robert E. Park and Ernest W. Burgess, The City (Chicago, Ill.: University of Chicago Press, 1925), p. 51

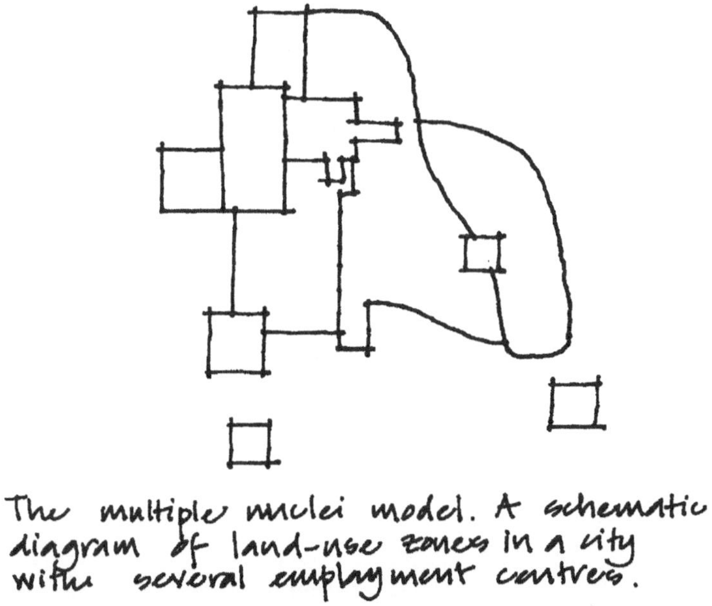

6.2 *Multiple nuclei city.* Source: Adapted from Chauncey D. Harris and Edward L. Ullman, "The nature of cities," Annals of the American Academy of Political and Social Science, 242 (1945): 13

they are often of similar importance. The residents choose their dwelling not on the basis of transportation to the CBD but rather on access to the particular center or centers where they work, shop, seek services or are entertained (Pederson, 1980: 25).

A third form of city development encompasses elements from the above two models. Pederson (1980) termed it the *sector model* (Figure 6.3). In the early stages of the development of the concentric model, the rich tended to congregate near the CBD, pushing land prices beyond the reach of others. This led to the development of routes out of the CBD which opened up the hinterland for dwelling, giving birth to the sector model. In this model, radial routes out of the CBD assume greater importance. As the route network developed, they attracted the rich to areas immediately surrounding these routes. The first mechanical transportation routes generally served the rich, pulling them out of the congested central area to places near the radial transportation lines. As more routes were added, radiating in different directions from the CBD, they served less-affluent people, each new line tending to serve lower-income people than the line which came before it.

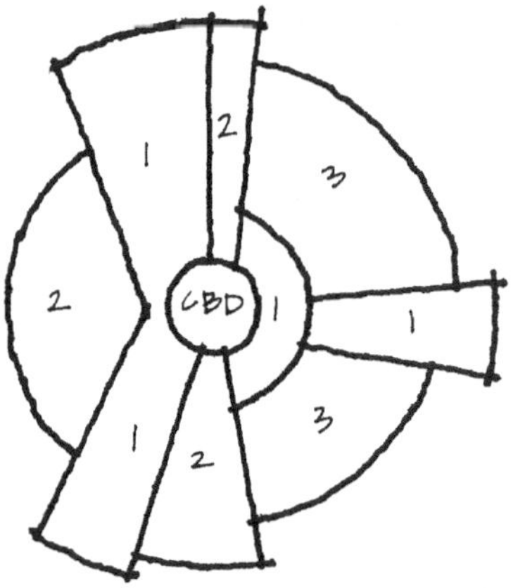

The sector model. Areas numbered 1, lower-income residential; areas numbered 2, high-income residential; areas numbered 3, medium-income residential.

6.3 *Sector model.*
Source: As Figure 6.2

6.2 Character of cities and their transport

A number of characteristics of cities are particularly relevant to transport plans and policies:

1. Population density
2. Layout of major streets, avenues and expressways, varying from a radial design that funnels movement toward the center, to a cobweb or grid that permits movement from everywhere to everywhere.
3. Micro-level features that influence local traffic circulation and traffic safety.
4. The visual form; that is, the apparent clarity or "legibility" of the city, reflecting the degree to which people receive orientation or are disoriented by the street pattern, landmarks, and other characteristics of the city.
5. Historical, recreational, and aesthetic environments that are to be preserved or created.
6. The financial situation of the city (see Wright, 1992: 145–146).

From the beginning of civilization until the first decades of the twentieth century, personal travel in cities was predominantly done by walking, occasionally supplemented with pack or draft animals. This applied to both personal travel and the movement of household, commercial, and industrial goods. The street was an extension of the urbanite's home, a place where friends met casually or by design, conversations were held, business transacted (Wright, 1992: 146).

Most important cities were located on or near a river front, sea port, or lake shore that, in addition to furnishing water, greatly facilitated commerce with other cities and countries. The problems with land transportation and the need for break-bulk operations encouraged the location of industry, commerce, and services in the immediate surroundings, providing an economic base for population growth. The larger cities of most countries are still found on such sites. Cities had to be compact, since slow and expensive horse and human power were the only means of transporting goods away from the rail stations and docks. In many European cities, the majority of the population lived within the cramped areas circumscribed by city walls.

The waterside location and other topographical features of cities have often been responsible for irregular street patterns. But, with walking as the predominant urban transport mode and the limited range of pedestrian traffic, surveyors, city planners and others attempted to reduce the distances between all points of the city and maximize the legibility of street patterns by introducing a grid pattern.

The reliance on walking required a high degree of compactness in cities. Walking more than a kilometer or two from home to downtown made trade and work difficult, and most cities were circumscribed by such a radius. This led to the enormous compaction of cities, with population densities as high as three times above the most densely populated contemporary cities. (In the early nineteenth century, large American cities had population densities above 30,000 persons/sq. km, about twice as high as Tokyo's present density [Wright, 1992: 148].) Such large concentrations of people, together with a lack of vaccines, potable water and sewerage systems, made the city a fetid and unhealthy place to live.

It is at this point that the electric streetcars made their appearance and made possible the expansion of the city area. Although bicycles made their appearance around this time too, there was no possibility for them to compete with electric street cars, on account of speed, carrying capacity, as well as traffic dangers (see Wright, 1992).

Landuse and transport

In general we can state that six variables of city planning affect urban transportation patterns:

1. *Density.* Higher densities decrease the number of trips taken and the percentage of auto and total energy consumed, but decrease speed and may increase trip length. Studies generally agree that mode split and travel speed/trip generation is highly correlated.
2. *Activity mix.* Studies show a weak link between land-use mix within specific areas and travel patterns in those areas.
3. *Jobs Decentralization.* The decentralization of jobs has a more complicated relationship to transportation demand. There is a clear trade-off between trip length and mode split as jobs decentralize to the suburbs: trips tend to be shorter but more often by automobile, although if jobs greatly exceed housing, trips are longer but more often involve ridesharing. The net impact of job decentralization upon travel and energy use is uncertain.
4. *Transportation provision.* Not surprisingly, higher provision of transit is associated with lower percentage of trips by auto, while higher provisions for automobile are associated with higher percentage of trips by auto. The direction of causality (which causes which, provision of transportation or travel pattern?) is uncertain.
5. *Overall city structure.* Studies are generally inconclusive on this aspect. Most studies conclude that a polycentric city form results in fewer trips, shorter trips, and less total energy use than either a monocentric or dispersed city form. Monocentric form, on the other hand, seems to result in the lowest percentage of trips by auto, while the dispersed city form experiences the highest percentage of trips by auto. There appears to be a trade-off between trip length and mode split.
6. *City size.* As city size increases, trip length and energy use

increase, but the relationship between size and mode split is uncertain, depending on density and the provision of transit. Similarly, as distance from the CBD increases, trip length, speed, and energy use increase, as does the percentage of trips by auto.

6.3 Transport ills of contemporary cities

Although traffic ills differ from city to city we can detect some commonalities among several urban conglomerations, particularly in the tropical world. These common factors have to do with urban structure or form.

The standard assumptions and popular beliefs about relationships between urban structure and travel patterns – in particular, that dense development results in less total travel by automobile and thus less energy use – are generally supported by most urban transportation studies, but not by all (Handy, 1992).

A detailed survey of the transportation ills of contemporary tropical cities is beyond the scope of this book. However, we would suggest that what lies at the heart of tropical urban transportation ills are two issues: the cost of mechanized transportation and energy equity. These two offer excellent starting points for further discussions.

Public costs of motorized urban transportation

Manning (1984) listed five possible social costs that are typically unpaid or underpaid by motorists:

1. The costs of road construction and maintenance.
2. Collision costs.
3. Environmental costs.
4. Subsidies to fossil fuels.
5. Cost to those not owning or having access to motor vehicles (Manning, 1984: 100).

Cost of motorways

While the costs of labor, materials and inflation rates play important parts, the most significant factor in road construction costs, at least in the urban areas, is the value of land. Manning (1984) estimated that up to 80 percent of urban road cost is tied to land costs. This can be a significant factor in high-density, Third World cities where land prices are typically higher than those in First-World cities (Gibbons, 1990).

The initial construction as well as the maintenance costs of motorways are higher irrespective of whether users are "public" or "private." They are also equally expensive in both developed and developing countries. Thus, railroad systems in urban Brazil cost

US$20–60 million/km (in 1990 values), while a similar project in Canada also costs roughly the same amount (Wright, 1992). The cost for highways is typically lower than for railroads, but, if land acquisition is involved, it, too, can approach the above figures.

Bus systems can be relatively cheaper, with the capital costs of buses being the largest item. However, in the urban Third World, this too can be enormous, as these systems usually have to be imported, paying hard currencies.

The operational and maintenance costs of motorways are also considerable. Wright (1992) found that in Brazil, upgrading an existing dilapidated roadway costs almost as much as constructing a new roadway.

Collision costs

In most countries, public expenses generated by traffic accidents are paid by taxpayers. The risk of personal injury or worse is not equally distributed among all segments of the population. The risk of collision is also unequally distributed among motorized transport modes (Table 6.3). Motorcycles pose the highest risk.

Considering the smaller number of motorized vehicles in the Third World, death rate per vehicle is higher in the developing world. For example, Brazil, with almost half of the population of the US, posted double the amount of accidental deaths caused by motor vehicles. More than two-thirds of the hospital beds in trauma and orthopedic wards in Brazil are occupied by accident victims (Wright, 1992: 89).

Buses are typically safer due to their size. Trains, too, are safer because of their exclusive tracks and elaborate safety network. However, a single train accident can be of disastrous proportions. The same is true for scheduled airlines.

Environmental costs

Transportation plays an important role in a country's energy consumption. Although the exact quantities vary from rich to poor countries, the

Table 6.3 Death rates for some motorized transport in the United States

Transport modes	*Deaths per 100 million passenger–km.*	*Risk relative to transit bus*
Motor cycle	25.5	4270.8
Passenger car	0.600	100.6
Passenger train	0.082	13.8
Scheduled airline	0.048	8.0
Intercity bus	0.044	4.6
Transit bus	0.006	1.0

Source: Wright (1992:104, Table 5)

pattern remains the same. Thus, while an industrialized country like the US spends 21 percent of its commercial energy consumption in transportation (Erley and Mossena, 1980), a Third World nation like Sri Lanka spends 16.4 percent (Wijemanne, 1989). It is likely that the bulk of this transportation energy is spent in the urban sector.

Thermo-electric energy used by all mechanized modes of transport almost always waste a great portion of the energy. Only about 10.3 percent of energy produced at a power plant is actually used by an electric vehicle. A gasoline-powered car uses only about 7.1 percent of the energy delivered to its fuel tank (Salihi, 1973). If energy losses in the mining and transporting of coal and crude oil are also included, these lower values could be reduced even further.

Within such lower-energy conversion rates, motorized vehicles show a remarkable degree of variance. For example, an intercity bus in a US urban street is about four times more energy-efficient (on the basis of passenger km traveled) than an airplane traveling on a certified route (Table 6.4). These differences result mainly from passenger capacities and not so much from energy-conversion technological differences between the modes.

Improvements in energy efficiency are possible if load factors can be increased. (Even a compact car carrying four passengers and doing 31 m.p.g. [13.1 km.p.l.] generates a fuel efficiency of 124 passenger miles/gallon [52.5 passenger km/lit], which is higher than many city buses in US cities.) It is also possible to conclude that cities relying heavily on least-energy-efficient modes (private automobiles) can be expected to waste more energy on the whole.

Wasted energy always leads to air pollution. And the air pollution trends in the megacities of the world are anything but encouraging. In the mid-1980s, for example, the Global Environment Monitoring System (GEMS) network, linked to the United Nations Environment Program and the World Health Organization (WHO), had information on

Table 6.4 Energy use by passenger vehicles in the US, 1985

Mode	*Load factor (passenger km/ vehicle km)*	*Relative energy use (Intercity bus = 1)*
Intercity bus	41.8	1.00
Intercity rail	19.1	2.12
Commuter rail	35.6	2.19
Transit bus	12.7	2.43
Transit rail	23.1	2.77
Automobile	1.7	3.20
Air-certified route	89.3	3.82

Source: Wright (1992: 49, table 4)

Note: passenger km = No. of Passengers x km traveled; vehicle km = vehicle x km traveled sulfur dioxide concentrations for 54 cities worldwide, and half of these cities had pollution levels near to or exceeding the maxima WHO considered tolerable. For suspended particulate matter, 41 cities were monitored, and 37 had borderline or excessive levels (French 1990: 9–11).

The metropolitan area of Mexico City is an especially interesting case, since it may be both the world's most populous urban area and its most polluted (Wright, 1992). There, the shares in total emissions of transport vehicles and (in parentheses) some percentages for private cars are: 55 percent (31 percent) of nitrous oxides; 83 percent (48 percent) of hydrocarbons; 12 percent of sulfur dioxide; 12 percent of particulates; 98 percent (67 percent) of carbon monoxide; and 100 percent (80 percent) of lead (Wright, 1992).

Transport-generated pollution in the form of ozone can be highly visible. Ozone is also known as smog and is not related to the ozone in the stratosphere, which protects the earth from excessive exposure to ultraviolet rays. Ozone is not emitted directly by vehicles; it is formed by subsequent complex reactions of sunlight with emissions of nitrous oxide, hydrocarbons and, to some extent, carbon monoxide. Ozone is thus formed only in the presence of sunlight, especially intense sunlight.

Airborne pollutants are also one of the major causes of damage to buildings, historical monuments, home furnishings, clothing, vehicles, crops, forests, lakes, and water supplies. Some of this damage has been reasonably well publicized, such as the Black Forest in Germany, the Acropolis in Greece, the Taj Mahal in India, and many ancient Italian and Egyptian monuments. However, the less-publicized acidification of lakes and water supplies and damage to crops and other vegetation may already represent a major biological crisis. Half of Canada's eastern lakes are acid-sensitive and 150,000 of them suffer biological damage, while air pollution has reduced crop yields in countries as diverse as the United States and China (French 1990: 17–24).

Subsidies to fossil fuels

Fossil fuel remains heavily subsidized, not only in the United States but also in Third World countries like Sri Lanka (Wijemanne, 1989). Oil prices, which went up immediately following the two oil crises of the 1970s (1973 and 1979), have remained low until recently, but are picking up again. Depending on the price fluctuations, poor countries and rich alike subsidize oil so that what motorists pay for use is only a fraction of the actual cost. The situation is the worst in non-oil-producing developing nations which have to import crude oil using their meager reserves of hard currencies.

Cost to the poor

Even if all motorists pay their actual costs in terms of road construction, operation and maintenance, as well as pay their fair share of fuel taxes, there will still be a segment of the population unable or unwilling to drive. There are poor who cannot afford a vehicle, those with physical and mental handicaps debarred from driving, minors, and those who choose not to drive. Ironically, motorized transport, that was supposed to bring freedom to all, has actually enslaved all to its dictates.

The case of inequity caused by such discrimination will be discussed later, but we must analyze the case of one car-owning household even in wealthy cities. Morris (1981) pointed out that the increase in one-person car use for journeys to work suggests that one-car households are effectively no-car households during the work week. This is also reflected in the number of trips households make in chauffeuring non-driving members.

Inequity in urban transport

Even if all the hidden costs are paid by motorists, mechanized transport discriminates against those who cannot afford to or don't want to drive – in three ways:

1. Mechanized transport creates distances only it can cover: those who rely on non-mechanized modes are left out.
2. Jobs follow mechanized modes, leaving those who cannot afford those modes chronically jobless.
3. In poor countries not blessed with fossil fuel, the social costs of importing as well as cleaning up the environmental degradation left behind by the mechanized modes of transport are burdened on the non-user majority.

There is also the distinction of access vs. mobility: greater reliance on automobiles seems to have given more and more people greater and greater mobility, but that mobility comes at a price of greater distances to be covered (see Lowe, 1994).

The social and environmental costs associated with increasing mechanization of transportation, particularly urban transportation, is also unsustainable in the long term. As mentioned before, mechanized transport requires large subsidies to build, operate and maintain which even the richer nations are finding difficult. The greater contribution of mechanized transport to air quality in urban areas is also a serious concern.

Unsustainability of mechanized urban transportation is amplified by the modal mix: higher dependence on low energy performers will lead to unsustainable mechanized transport for the city as a whole. Table 6.5 shows the relative energy use by passenger vehicles in the

Table 6.5 Fuel efficiency for US transport vehicles

Mode	*Energy use passenger m.p.g.*	*Energy use available seats m.p.g.*
Intracity bus	75–259	217
Intercity train	82–252	450
Suburban train	200	400
Commuter train	100	200
Automobile	25	120
Airplane	14	68

Source: Wright (1992, table 6)

Notes: A Passenger mile is transporting one passenger over one mile; available seats mile is transporting x number of seats over one mile

United States. Single-passenger, automobile-dominated cities will have the most inefficient urban transportation system.

Greater reliance on mechanized modes of transport for urban traffic is also self-defeating in that these modes typically have lower urban passenger capacities than human-powered modes like walking and cycling. A walkway can carry more than double the amount of people in cities than a 30-seat bus: six times more than fully occupied cars running on wide streets.

A combination of greater reliance on inefficient modes and lack of facilities for human-powered modes, leads to energy inefficiency for the city as a whole. However, individual vehicles may achieve greater efficiencies. Newman and Kenworthy (1988) showed this in a study conducted in Perth, Australia. As more and more people lived and traveled greater distances from the central city, the fuel use per capita increased though their vehicular kilometers per gallon were improving. Technologies notwithstanding, a sprawling city will inevitably be energy-wasteful, and therefore unsustainable.

However, the pollution that comes from a greater reliance on mechanized modes of transportation is not solely physical. Important as it may be, air pollution emanating from mechanized transportation can be controlled, though at a price. The greater evil seems a social one. Illich called it a "dependence on energy that brings about social inertia" (1974: 6).

Even if non-polluting power were feasible and abundant, "the use of energy on a massive scale acts on society like a drug that is physically harmless but psychically enslaving. A community can choose between … maintaining its addiction or kicking it in painful cramps, but no society is at once autonomously active and hooked on progressively larger numbers of energy slaves" (Illich, 1974: 6).

We could therefore summarize the urban transport ills of tropical cities as follows:

1. Current policies related to urban transportation have led to transport inequity.

Table 6.6 Transport capacities and average speeds

Mode	*Capacity (persons/meter/hr)*	*Speed (km.p.h.)*
Non-motorized modes		
Walkway	3,609	3
Bikeway	1,476	13
Mixed Traffic, city street		
Cars with 4 occupants	381	24
Minibus (10 passengers)	492	16
Bus (30 passengers)	982	14
Expressway		
Cars with 4 occupants	2,362	64
Minibus (10 passengers)	3,937	64
Bus (40 passengers)	6,562	64
High-capacity urban rail		
Metro (22,500 pass./line hr)	5,577	34
Metro (40,000 pass./ line hr)	9,915	34

Source: Wright (1992)

Note: A wider street (13.4 m or wider) will increase the capacity for cars by about 10 percent

2. Equity in urban transportation is achieved only if maximum limits are set for per capita energy use.
3. Such an approach has implications for urban transportation policy and urban design.
4. This approach is also compatible with the goals of environment-conscious urban design.
5. Sustainable cities can grow out of practices that seek to minimize the reliance on mechanical/fossil-fuel power.

6.4 Transportation alternatives for environment-conscious urban development

Essential considerations for equitable and sustainable urban transportation

A review of principles and strategies for equitable and sustainable urban transportation appropriate for Third World cities is given in Emmanuel (2001). The following section summarizes the main findings.

Ceiling on per capita energy use

It is therefore suggested that sustainable and equitable urban transportation strategy be developed on the following basis:

- Beyond a critical speed, the higher the speed, the longer the time spent on travel (higher speeds can only be possible when travel distances are long).
- Speed also nibbles away lifetime, multiplies trips ("because it is possible!"), increases time spent on preparing for the trip, increases danger to life and limb (and therefore increases the cost of protection).
- Thus a cap on urban transport speed (i.e. energy use) is a necessary condition for decreasing the total time spent on travel and to improve urban quality of life.

Energy-efficient traffic vs. energy-efficient cities

Energy-efficient traffic (fuel-efficient traffic)
- High ratio of road space to population.
- High average speeds.
- High kilometers per litre.
- But each urban dweller travels longer distances, thus burning more fuel.

Energy-efficient cities
- Lower ratio of road space to population.
- Low average speeds.
- Low kilometers per liter.
- But shorter travel distances for everyone, thus less burning of fuel.

Climate and urban transportation

The problem of walking and riding bicycles in hot-humid cities is that the high air temperature, coupled with high relative humidity and traffic congestion, is enervating (Ondaatje, 1979). Therefore, shaded pathways exclusively designated for pedestrians and bicycles can be provided to encourage these modes of transport.

- Compact cities, easy to cool.
- Higher building densities, more open spaces.
- Possibility of speed separation, varying cooling enhancement possibilities, varying orientation possibilities.

Environmentally conducive urban transportation goals

- Less air pollution.

- Less noise.
- Less separation of urban space.
- More social interaction.

Recovery of the "urban commons"

- Importance of "urban commons" to climate-conscious urban design.
- Energy-efficient cities and "urban commons."

In summary, climate-sensitive urban design in the hot-humid context would entail at least the following:

1. Incentives for energy-efficient site development.
2. Greater diversity of land-use development.
3. Encouragement of efficient transport forms.

6.5 Design guidelines

Urban design actions can then enhance the above-identified objectives, particularly the last one. As Wright (1992: 245) says, a workable urban transport approach should be that "the government . . . do what the private sector cannot do or cannot do as efficiently as government; other tasks should stay in the private sector."

Create "Erf"-like environments in each neighborhood

An "Erf" is a Dutch urban design approach where streets are treated like backyard gardens that are mostly pedestrian-oriented (Figure 6.4). Cars are usually not allowed, except those of the residents. The right of way is purposely made to meander, incorporating many traffic bumps, thus discouraging mechanized modes of traffic.

All commercial development within neighborhoods to have arcades

The tropical climate makes walking a tedious exercise unless solar radiation is sufficiently shielded (Figure 6.5). Arcades could act as an extension of the "commons," where not only walking would be pleasant but various allied activities like sidewalk hawking could be possible.

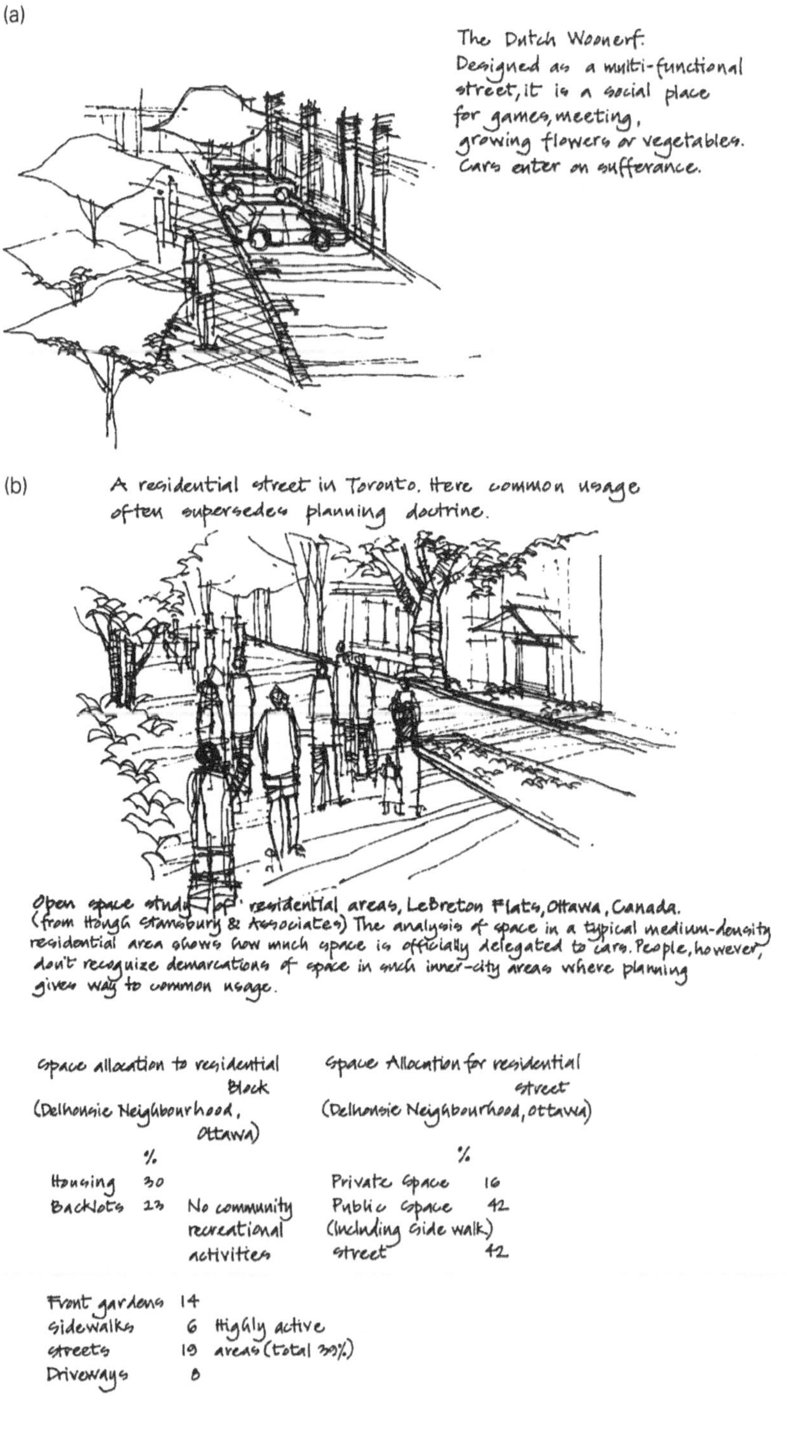

6.4 *"Erf" in urban design.*
Source: Hough (1984)

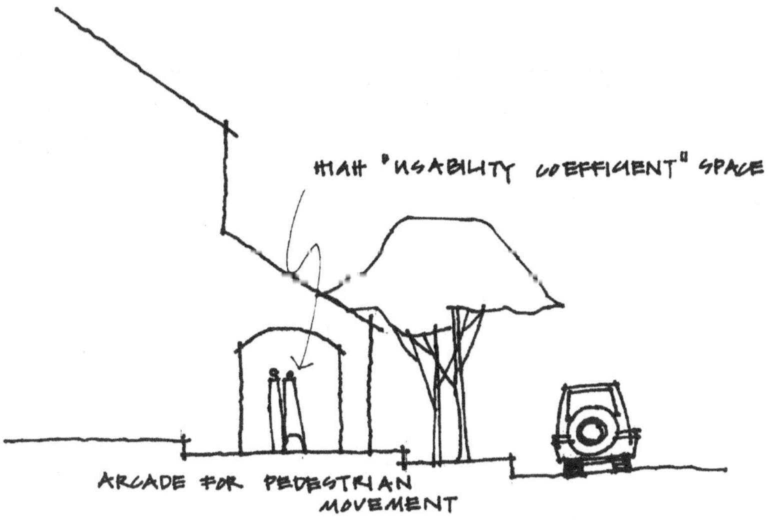

6.5 *Commercial development and transportation*

Prevent parceling of neighborhood land into small plots

Smaller plots use land inefficiently, posing a tangible threat to the "commons," and reduce green areas (thus also the "walk-ability" of a neighborhood). By acquiring land that comes up for sale, urban local authorities could direct development towards medium-rise, mixed-use structures, freeing up green space that can then be turned into continuous pedestrian networks.

Avoid single-function land use within neighborhoods

Single-function land uses like residential, commercial, etc., segregate urban activity and perpetuate traffic. By opting for more "mixed-residential" land use the need for traffic can be reduced. Activities that can be allowed are housing of different scales, light-commercial activities, offices, and service-sector functions that can be compatible with residential use and also generate some jobs.

North–South running streets should be treated as pedestrian links

Since sun angles in the tropics are high, solar control becomes a difficult issue. By making North/south running streets narrow and thus

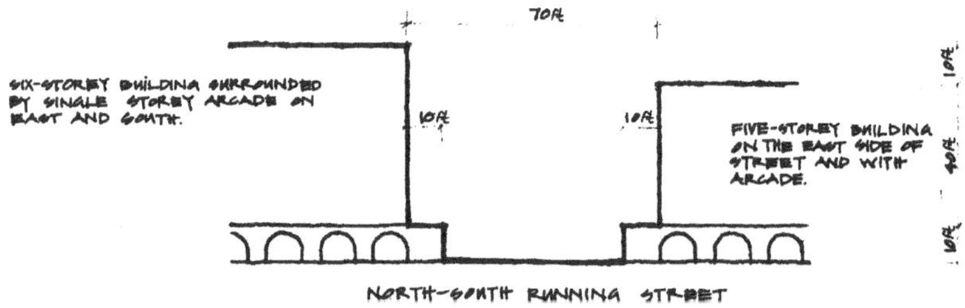

6.6 *Strategies for north–south streets*

shady, pedestrian movement within neighborhoods can be encouraged (Figure 6.6).

East–West running streets could be erf-like environments

Since it is relatively easier to ward off northern/southern sun in the equatorial tropics, streets in this direction could be wider (Figure 6.7). A varied environment, allowing arcade development, bicycles, some public transit options, etc., could be created.

Permit the use of "shared taxis" on major traffic arteries

Variously known as jeepneys, jitneys, peseros, puerpestos, etc., shared taxis can reduce the load on public transit. These fixed route vehicles are more flexible (though comparatively expensive) and can be permitted within neighborhoods as well.

Adapt established neighborhoods for walking and cycling

Once the necessary policy framework is in place, each neighborhood unit can be retrofitted to facilitate walking and the use of bicycles.

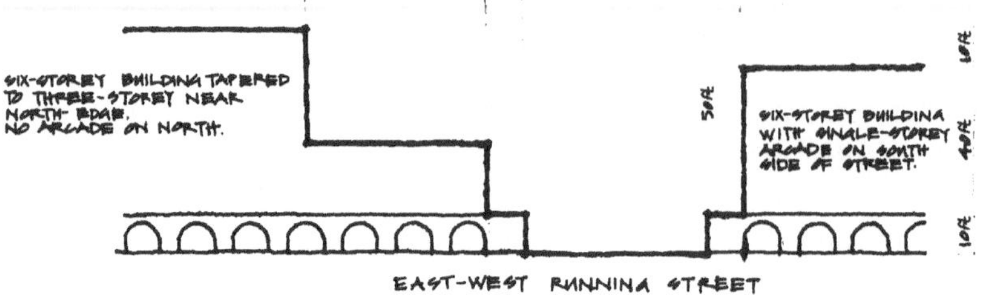

6.7 *Strategies for east–west streets*

Specific strategies in this direction would include widening sidewalks, automobile slowing-down measures (not more than 24 km/h), street diverters, increased sidewalk area at pedestrian crossings, traffic circle, etc. (Figure 6.8).

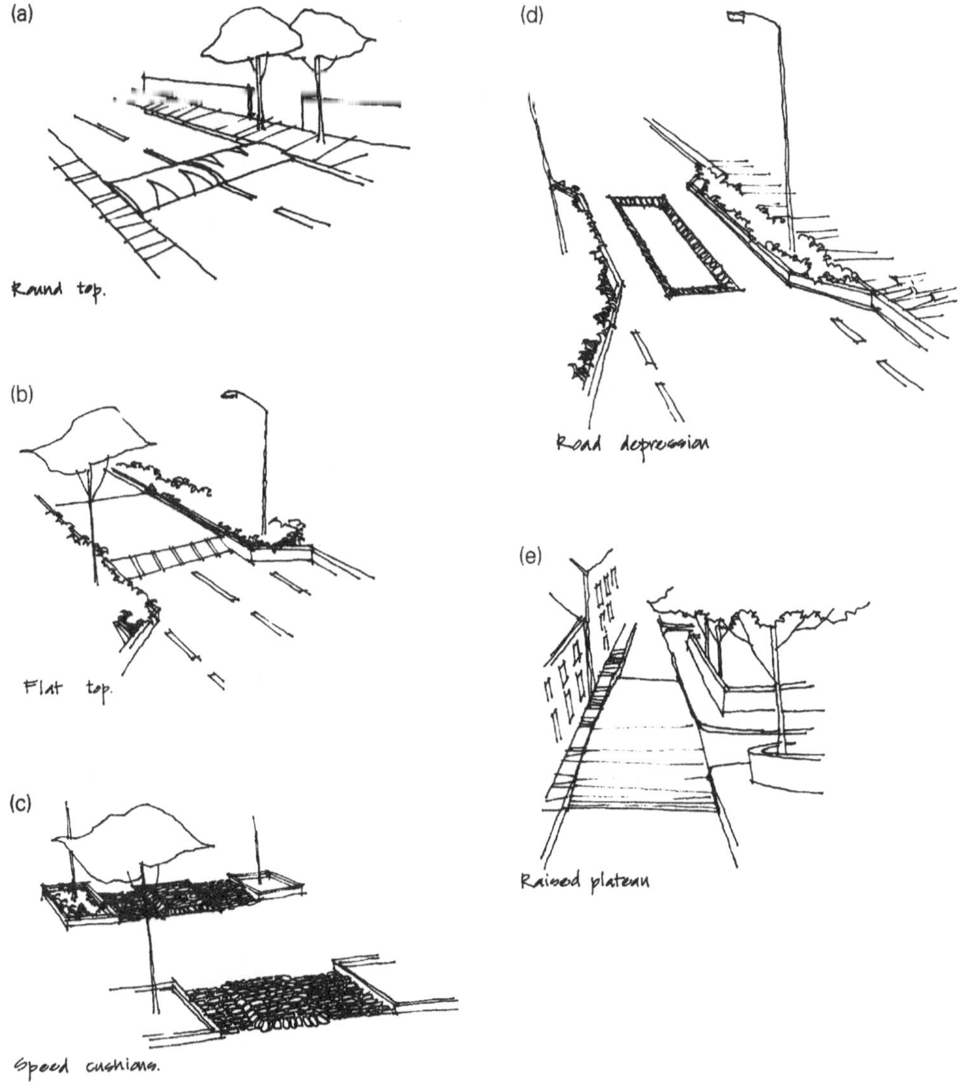

6.8 *Adapting existing neighborhoods to traffic calming*

6.6 Conclusions

One of the significant features of the urban design options listed above is the use of existing facilities. Very little new infrastructural inputs are necessary. Where needed, these will generally be in the form of building arcades and other improvements for pedestrianizing neighborhoods which can be accomplished at a fraction of the cost of building and maintaining freeways. The aim is to recover the public realm, or the commons, thus empowering people to choose transit modes, including walking. Only such an attempt would truly democratize urban transport.

The objectives and strategies listed here are by no means exhaustive. The point is to show that such an approach is possible even for poor, Third World cities. One objection, however, could be that traffic resulting from the above approach would be slower. That is precisely the point, though it is by no means certain. Also, slower traffic allows the socially disadvantaged to compete realistically. Pedestrianization and other such humanizing attempts at city levels have long been carried out by progressive western cities. There is a perception among some that such attempts are possible only in the developed west. It is therefore necessary that a comparable approach be initiated in the Third World as well. It is hoped that on the basis of the general directions outlined above, more detailed strategies can now be developed to direct urban growth in tropical cities in a sustainable manner.

References

Correa, C. (1989). *The New Landscape: Urbanization in the Third World*, London: Butterworth Architecture.

Emmanuel, R. (2001). "Strategies for equity and sustainability in urban transportation in Third World mega-cities," *Built-Environment Sri Lanka*, 2(1): 20–34.

Erley, D. and D. Mossena (1980). *Energy Conserving Development Regulations: Current Practice*, Planning Advisory Service (PAS) Report No. 352, Chicago, Ill.: American Planning Association.

French, H.F. (1990). *Cleaning the Air: A Global Agenda*, Worldwatch Institute Paper No. 94, Washington, D.C.: Worldwatch Institute.

Gibbons, S. (1990). "Mixed land-use and multi-modal transport: the Kanpur planning environment," (quoted by Wright, 1992, op. cit.).

Handy. S. (1992). *How Land Use Patterns Affect Travel Patterns: A Bibliography*, Chicago, Ill.: Council of Planning Librarians.

Illich, I. (1974). *Energy and Equity*, London: Calder and Boyars.

Lowe, M.D. (1994). "Reinventing transport," in L. Brown *et al.*, (eds), *State of the World 1994*, New York: W.W. Norton, pp. 81–98.

Manning, I. (1984). *Beyond Walking Distance: The Gains from Speed in Australian Urban Travel*, Canberra: Australian National University, Urban Research Unit.

Morris, J.M. (1981). "Urban public transport," in P.N. Troy (ed.), *Equity in the City*, Sydney: George Allen and Unwin, pp. 21–49.

Newman, P.W.G. and J.R. Kenworthy (1988). "The transport energy trade-off: fuel-efficient traffic versus fuel-efficient cities," *Transportation Research, Part A*, 22A (May): 163–174.

Ondaatje, M. (1979). *Running in the Family*, New York: W.W. Norton.

Pederson, E.O. (1980). *Transportation in Cities*, New York: Pergamon Press.

Salihi, J.T. (1973). "Energy requirements for electric cars and their impact on electric power generation and distribution systems," *IEEE Transactions*, 1A-9(5): 516–533.

Wijemanne, E.N. (1989). *Energy Pricing Options for Sri Lanka*, New Delhi: International Labor Organization/Asian Regional Team for Employment Promotion.

Wright, C.A. (1992). *Fast Wheels, Slow Traffic: Urban Transport Choices*, Philadelphia, Pa.: Temple University Press.

Index

Italic page numbers indicate figures and illustrations not included in the text page range.